Making Deep Maps

This book explores how we create deep maps, delving into the development of methods and approaches that move beyond standard two-dimensional cartography.

Deep mapping offers a more detailed exploration of the world we inhabit. Moving from concept to practice, this book addresses how we make deep maps. It explores what methods are available, what technologies and approaches are favorable when designing deep maps, and what lessons assist the practitioner during their construction. This book aims to create an open-ended way in which to understand complex problems through multiple perspectives, while providing a means to represent the physical properties of the real world and to respond to the needs of contemporary scholarship. With contributions from leading experts in the spatial humanities, chapters focus on the linked layers of quantitative and qualitative data, maps, photographs, images, and sound that offer a dynamic view of past and present worlds.

This innovative book is the first to offer these insights on the construction of deep maps. It will be a key point of reference for students and scholars in the digital and spatial humanities, geographers, cartographers, and computer scientists who work on spatiality, sensory experience, and perceptual learning.

David J. Bodenhamer is (founding) executive director of the Polis Center and Professor of History and Informatics at Indiana University–Purdue University, Indianapolis. He writes widely in the field of spatial humanities and is coeditor of the Routledge Series on Spatial Humanities.

John Corrigan is the Lucius Moody Bristol Distinguished Professor of Religion and Professor of History at Florida State University. His previous coauthored books with Routledge include *Religion in America* and *Jews, Christians Muslims: A Comparative Introduction to Monotheistic Religions*. He is coeditor of the Routledge Series on Spatial Humanities.

Trevor M. Harris is Eberly Distinguished Professor of Geography Emeritus at West Virginia University. He is coeditor of the Routledge Series on Spatial Humanities. His research focuses on GISc, immersive and virtual GIS; Spatial Humanities; Deep Mapping; Exploratory Spatial Data Analysis; and Critical and Participatory GIS.

Routledge Spatial Humanities Series

Series Editors: David Bodenhamer
Indiana University Purdue University, USA

John Corrigan
The Florida State University, USA

Trevor M. Harris
West Virginia University, USA

This series explores how considerations of space and place, both geographical and metaphorical, are shaping innovative scholarship in the humanities. It represents a bridging of disciplines, with history, archaeology, literary studies, religious studies, and cultural studies, among others, now taking up theories and approaches associated most often with geography and the social sciences. Books in the series explore theoretical, methodological, technological, and applied issues, such as deep mapping, immersive geographies, experiential and emotional spatial worlds, and time-space linkages.

Forthcoming titles

Making Deep Maps
Foundations, Approaches, and Methods
Edited by David J. Bodenhamer, John Corrigan and Trevor M. Harris

Making Deep Maps

Foundations, Approaches, and Methods

Edited by David J. Bodenhamer, John Corrigan and Trevor M. Harris

Routledge
Taylor & Francis Group

LONDON AND NEW YORK

First published 2022
by Routledge
2 Park Square, Milton Park, Abingdon, Oxon OX14 4RN

and by Routledge
605 Third Avenue, New York, NY 10158

Routledge is an imprint of the Taylor & Francis Group, an informa business

British Library Cataloguing-in-Publication Data
A catalogue record for this book is available from the British Library

Library of Congress Cataloging-in-Publication Data
A catalog record for this book has been requested

ISBN: 978-0-367-74383-3 (hbk)
ISBN: 978-0-367-74385-7 (pbk)
ISBN: 978-0-367-74384-0 (ebk)

DOI: 10.4324/9780367743840

Typeset in Times New Roman
by Apex CoVantage, LLC

Contents

List of figures	vii
List of contributors	x
Preface	xiii

1 **The varieties of deep maps** 1
DAVID J. BODENHAMER

2 **The art of deep mapping** 17
DENIS WOOD

3 **Designing for mysterious encounter: three scales of integration in deep mapmaking** 38
NICHOLAS BAUCH

4 **Spatializing text for deep mapping** 50
MAY YUAN

5 **Representational issues in deep mapping: peeling the "poetic and positivistic" from the western geosophical onion** 65
CHARLES TRAVIS

6 **Indigenous deep mapping: a conceptual and representational analysis of space in Mesoamerica and New Spain** 78
PATRICIA MURRIETA-FLORES, MARIANA FAVILA-VÁZQUEZ, AND ABAN FLORES-MORÁN

7 **Deep mapping the lived world: immersive geographies, agency, and the virtual umwelt** 112
TREVOR M. HARRIS

8 **Navigating through narrative**   132
   STEPHEN ROBERTSON AND LINCOLN A. MULLEN

9 **Cultural heritage institutions and deep maps**   148
   MIA RIDGE

10 **The inexactitude of science: deep mapping and scholarship**   162
   JOHN CORRIGAN

11 **Convergence: GIScience turns to the spatial humanities**   175
   KAREN KEMP

   Index   181

Figures

2.1   The map inside the Brattleboro Museum and Art Center's
      exhibition guide, *Town Treasures 8: The Portrait* (1990)   18
2.2   Second State Overlook Impression Organization was the
      sixth in the series of maps of Cleveland Heights   21
2.3   This is the map of Overlook I made in 1976 showing the
      location of my paper routes   22
2.4   Mark Salling's map of his paper route and environs   23
2.5   One of the maps from my master's thesis, Fleeting Glimpses   24
2.6   The color of the leaves in Boylan Heights in the fall, 1982   26
2.7   The map of the streetlights Carter Crawford drew with my
      pochoir brush   27
2.8   The map of my living room at 435 Cutler Street   29
2.9   The place of our house (with the dot on it) in Boylan Heights   30
2.10 The place of Boylan Heights   31
2.11 This is a map of the "Absentee Landlords" living in Raleigh   32
2.12 This is the map I made of "Assessed Value" in 1982   33
2.13 Boylan Heights is at the intersection of these wave fronts being
      generated by the towers indicated   35
2.14 The wave fronts generated by the stations plotted in Figure 2.13 36
4.1   A word cloud representation of Bodenhamer et al. (2013):
      Deep Mapping and Spatial Humanities   53
4.2   A semantic word cloud of Bodenhamer et al. (2013): Deep
      Mapping and Spatial Humanities   54
4.3   IN-SPIRE stands for In Spatial Paradigm for Information
      Retrieval and Exploration   56
4.4   A general workflow for toponym resolution in text   59
5.1   Google Earth Mapping of GPO Dublin, Ireland with location
      of the death of Robert Patrick Heeney, a civilian killed while
      trying help a wounded man   69
5.2   Network map connecting location of civilian deaths (white and
      large circles) to locations of their homes (black dots) by latitude
      and longitude   70

5.3   2016 Commemoration Social Media Posts juxtaposed with
      Ground Zero (Black Ellipses) of 1916 Rising Deaths proximate
      to the General Post Office   71
5.4   Bloomsday "Sense of Place," Dublin 1954 and 2014   72
5.5   *Ulysses* Deep Map of Dublin   73
5.6   Odyssean and Dantean Mapping of Dublin, Ireland   73
6.1   Representation of Mount Tlaloc on page 22 of the Codex
      Borbonicus   83
6.2   (a) Detail of page 11 of the *Matrícula de Tributos*; (b) Detail
      of page 5 of the *Matrícula de Tributos*; (c) Detail of page 5
      of the Codex *Boturini*   84
6.3   Detail of page 14 of Codex *Borgia* depicting *Tepeyollotl* in front
      of a temple   85
6.4   Diagram showing the different multiscalar relations of the
      altepetl   86
6.5   The Relación Geográfica de Meztitlán map   89
6.6   The Relación Geográfica de Atengo-Mixquiahuala map   91
6.7   Tula is represented on the lower-right end of the map (identified
      with the gloss) while the toponym of Cerro Xicuco is shown on
      the right-hand side of the church.   95
6.8   The Relación Geográfica of Teozacoalco map   96
6.9   Detail of the starting point of the genealogies of the rulers of
      Teozacoalco and Tilantongo   97
6.10 Representation of the rulers of the Apoala Valley and its
      geographic and sacred landscape in the Zouche-Nuttall Codex,
      folio 37   98
6.11 Lord 2 Dog is represented as a Chichimeca warrior with bow
      and arrows in his hand while confronting the people
      of Teozacoalco   99
6.12 The Relación Geográfica of Cempoala map   100
6.13 The toponym of Cempoala   101
6.14 A member of the community presents an offering to the
      Lienzo de Petlacala   104
6.15 The *lienzo* is placed in the altar at the top of Chichitépetl Hill
      for the petition of rain   105
7.1   Rebecca Harding Davis's Wheeling, WV in the mid-nineteenth
      century   121
7.2   Wheeling mid-nineteenth-century streetscene   122
7.3   The virtual landscape of Blair, WV   124
7.4   Virtual Blair, WV   125
7.5   Nighttime lighting in a deep map   126
7.6   Nighttime and the placement of emergency telephones   127
8.1   *Mapping Occupation* is an example of a spatial narrative using
      ESRI's Story Map   137
8.2   *Year of the Riot* is a deep map of the 1935 Harlem Riot   139

8.3   *Year of the Riot* showing broken windows (left) and looted
      locations (right)   141
8.4   The initial overview at the beginning of the spatial narrative
      of the 1935 Harlem riot   142
8.5   The spatial narrative of the riot in Neatline zooms in and
      displays the relevant items from the deep map, accompanied
      by narrative text   143

Contributors

Nicholas Bauch is a research fellow at the Institute for Advanced Study at the University of Minnesota. He is an artist and a geographer who utilizes writing, printmaking, performance, and photography to critically embody landscapes. He holds an MFA in Visual Art from the University of Minnesota (2021) and a PhD in Geography (2010) from the University of California–Los Angeles, where he specialized in cultural and historical geography. He is author of *Enchanting the Desert* (2016, Stanford University Press), and *A Geography of Digestion* (2017, University of California Press). Prior to pursuing an MFA degree, he was founding faculty director of the Experimental Geography Studio at the University of Oklahoma. He has taught geography, photography, history, and language at six universities since 2004. Between 2012 and 2016, he was a scholar at Stanford University's digital humanities research center (www.geographystudio.org).

David J. Bodenhamer is Professor of History and Informatics and (founding) executive director of the Polis Center at Indiana University–Purdue University, Indianapolis (https://polis.iupui.edu). His research interests in the spatial humanities are in deep mapping and spatial narratives. He has published and lectured widely in spatial humanities and from 2007 to 2020 served as editor of *IJHAC: A Journal of the Digital Humanities*. He also is an expert in US constitutional and legal history, with six books and numerous journal articles.

John Corrigan is the Lucius Moody Bristol Distinguished Professor of Religion and Professor of History, and Distinguished Research Professor at Florida State University. His research is in the areas of American religious history, religion and emotion, religious violence, and the spatial humanities. His current research is on trauma and genocide. He is coeditor of the Routledge Series on Spatial Humanities.

Mariana Favila-Vázquez specializes in Mesoamerican archaeology and historical archaeology. She is a research associate of the project "Digging into Early Colonial Mexico: A large scale computational analysis of historical sources" and a visiting lecturer at the National School of Anthropology and History (ENAH) and at the National Autonomous University of Mexico (UNAM). Her research has focused on the Prehispanic navigation systems and landscapes, and studying historical cartography with spatial technologies. She is the author of the book *Veredas de mar de río. Navegación prehispánica y colonial en Los Tuxtlas, Veracruz.*

Aban Flores-Morán is a lecturer at the Art Department in the Center for Education for Foreigners (CEPE-UNAM). He specializes in art history and historical archaeology. He is also a visiting lecturer at the National School of Anthropology and History, the Autonomous University of San Luis Potosí (UASLP), the Hellenic Cultural Center, and the Postgraduate Program in Art History (UNAM). His research line focuses on the study of the transformations, continuities, and coexistences of the indigenous and western image during the sixteenth century in New Spain.

Trevor M. Harris is Eberly Distinguished Professor of Geography at West Virginia University. Dr Harris codirects the State GIS Technical Center and the Virtual Center for Spatial Humanities. He obtained his doctorate from the University of Hull, England. Dr Harris's research focuses on Geographic Information Science and includes Immersive Virtual Reality and GIS, 3D GIS, and 3D modeling; the Spatial Humanities; Deep Mapping; Exploratory Spatial Data Analysis; the Geospatial Semantic Web; and Critical and Participatory GIS.

Karen Kemp is Professor Emerita of Spatial Sciences at the University of Southern California. For several years before retirement, she taught full-time online in the USC Master's Program in GIScience and Technology from her home in Hawaii. She has been recognized globally for her major contributions to the growth and evolution of GIScience education.

Lincoln A. Mullen is an Associate Professor in the Department of History and Art History at George Mason University and the director of computational history at the Roy Rosenzweig Center for History and New Media. He works with digital history methods, especially mapping and text analysis, for the study of American religion. He is the author of *The Chance of Salvation: A History of Conversion in America*, and he is currently working on a digital monograph titled *America's Public Bible: A Commentary*. With John Turner, he is co-directing a project titled *American Religious Ecologies*, which will digitize and map the 1926 Census of Religious Bodies.

Patricia Murrieta-Flores is Co-Director of Digital Humanities and Senior Lecturer at the History Department of Lancaster University in the UK. She specializes in historical archaeology and has published widely on the application of computational methods for humanities research. Her research interests lie in Mesoamerican and Mexican colonial history and Artificial Intelligence methods for historical and archaeological research. She leads multiple funded interdisciplinary projects, including "Digging into Early Colonial Mexico: A large-scale computational analysis of 16th century historical sources" and "Unlocking the Colonial Archive: Harnessing Artificial Intelligence for Indigenous and Spanish American Collections."

Mia Ridge is the British Library's Digital Curator for Western Heritage Collections. A core member of the Library's Digital Scholarship team, she enables innovative research based on digital collections. She leads large- and small-scale projects, providing guidance and training on computational methods for historical collections. Her research is on aspects of human–computer interaction and digital cultural heritage, particularly on crowdsourcing in galleries, libraries, archives, and museums has an international reputation. Her edited volume, *Crowdsourcing Our Cultural Heritage* (Ashgate) was published in October 2014 and subsequently issued in paperback.

Stephen Robertson is a Professor in the Department of History and Art History at George Mason University, where from 2013 to 2019 he served as executive director of the Roy Rosenzweig Center for History and New Media. With colleagues at the University of Sydney, he created the award-winning *Digital Harlem*, a research tool that maps the complexity of everyday life in the 1920s. He is currently working on *Harlem in Disorder: A Spatial History of How Racial Violence Changed in 1935*, a multi-layered, hyperlinked digital publication created in Scalar that is under contract with Stanford University Press.

Charles Travis is Assistant Professor of Geography and GIS with the Department of History at the University of Texas, Arlington, and a Visiting Research Fellow with the School of Histories and Humanities at Trinity College Dublin. As an editorial board member of the journal *Literary Geography*, his book publications include: *Historical Geography, GIScience and Textual Analysis* (2020), *Abstract Machine: Humanities GIS* (2015) *History and GIS: Epistemologies, Reflections and Considerations* (2012) and *Literary Landscapes: Geographies of Irish Stories, 1929–1946* (2009). His work has also appeared in numerous professional journals.

Denis Wood is a geographer, independent scholar, and author of books on maps, their power and subjectivity, including the influential *The Power of Maps*, which originated as an exhibition Wood curated for the Smithsonian. More recent publications include *The Natures of Maps* with John Fels; *Rethinking the Power of Maps*; *Making Maps*, third edition, with John Krygier. *Everything Sings*; and *Weaponzing Maps* with Joe Bryan.

May Yuan is Ashbel Smith Professor of Geospatial Information Sciences and GIS PhD director in the School of Economic, Political, and Policy Sciences at the University of Texas at Dallas (UT-Dallas). Her research interest centers on developing approaches to represent geographic events and processes in GIS databases to support space–time query, analytics, and knowledge discovery.

Preface

Its title notwithstanding, this volume contains no instructions for making deep maps. No manuals, no algorithms. We wish it were that easy: plug here, play there, and *voila*, a deep map. Our aim, paradoxically, is both simpler yet more complex. It is to ground the discussion of deep mapping theoretically, to broaden its definition to encompass multiple expressions of deep maps, and especially to offer guidance and examples on the various methods and tools scholars are using to construct these complex products. We may not offer templates for a deep map, but scholars and students in a variety of disciplines will have a much better understanding of how to think about the opportunities and challenges this approach offers for their research.

Although numerous scholars have traced its provenance too simplistically to William Least Heat-Moon's book, *PrairyErth* (1991), fewer have understood the current interest in deep mapping as a response to the inadequacy of geographic information systems to serve the needs of humanists.[1] Maps are symbolic representations of an environment—usually, but not always, the physical world—that provide important knowledge for users. Early maps were encyclopedias that contained information about religion, government, economy, myths, and other social and cultural knowledge, as well the locations of places, features, boundaries, and routes. These maps were ill-suited for the modern world. Claims of empire, wars and military necessity, the emergence of transnational markets and industrial economies, new modes of transportation, the shift to widespread ownership and exchange of real property, and a host of other forces led to the carefully plotted, scientifically measured maps we use today. The arrival of geographic information systems (GIS) made convenient and economical what once was the province of experts and, with global positioning systems, made possible the navigation apps that are standard on our smartphones. But these developments also embodied a somewhat sterile, route-based view of the world that relied heavily upon an administrative, material, and physical view of reality.

Deep mapping embodies an acknowledgement that what we know about the world, even scientifically, is socially constructed. We each see the world differently, depending on our unique experiences and on the intellectual framework we use to make sense of them. The emergence of feminist geography in the 1970s helps us to understand this point. Led by such scholars as Linda McDowell and

Joni Seager, among a host of others, it became clear that ideas about gender, race, class, and other categories shaped the way people understood and navigated space. A woman's "place" was not merely a symbolic expression of separateness, it marked a real geography of difference that provided women with experiences and perspectives that did not correspond with dominant male views.[2] We understand that women's history, for example, is not a simple matter of identity politics in which we seek to resurrect female role models to stand alongside male role models. Rather, we now acknowledge that women's experiences are different from men's and often lead to a different understanding of such matters as power, relationships, causation, and meaning, among other things. Our maps should reveal these differences. We cannot freight our maps with every perspective, but we should insist that they reflect the relative indeterminacy of knowledge. The aim here is to remind us that the places we make and the events we chart are complex and filled not only with multiple views but also with deep contingencies. Although a powerful technology, GIS measures the world; deep maps help us interpret it.

Making deep maps, then, is more than a matter of applying GIS and related geospatial tools, no matter how central the technology may be in constructing the map itself. The aim is to embed the map with the evidence and perspectives we require to understand the problems and issues that interest us. To accomplish this result, we may have to push the map beyond its typical 2D expression into different forms, using a wide array of technologies and methods from linked data and geovisualization to augmented and virtual reality and immersive environments. We must insist that the map acknowledge the incompleteness and ambiguity of our data and be open to new information as we discover it. We must find ways to move the textual evidence that is most available to humanists into this new flexible, open-ended environment, and we must require that it allow us to traverse temporal and spatial scales, thereby making it more fluid and more malleable. Above all, the map must invite exploration, perhaps raising more questions than it provides answers.

The essays in this volume will help guide thinking about how best to accomplish these aims. The order of these contributions moves from the historical, conceptual, and theoretical underpinnings of deep mapping to its methods and its implications for humanities disciplines. Each essay may be read separately with profit. Collectively the essays are not designed to convey a single point of view but rather to reveal the rich variety of approaches that are emerging in this nascent field, which manifests itself in disciplines from archaeology and anthropology to geography, history, and literature, to name a few. Deep mapping is not yet sufficiently mature to claim that the essays in this volume will be encyclopedic, but they will offer grounding in theory and practice, as well as point to promising new directions. The notes for each chapter also offer ample evidence to the rich and expansive literature that has fed the impulse toward deep mapping.

One conclusion is inescapable: deep maps exist in multiple forms, as David J. Bodenhamer notes in his essay. The varieties of deep maps fit into five categories and maybe more, each designed for its own purpose. Also, as Denis Wood argues, deep mapping does not require technology, even though most authors in this volume use it effectively. What is essential, however, is a keen engagement with the world of memory, whether personal or evidentiary, as well as the world of matter. The approach depends on art as much as it does science, and it is inexact and oriented toward experimentation.

One of the most challenging problems for humanities researchers interested in spatial humanities is the use of textual evidence, which after all is a primary source of information for many questions that concern them. Here, historians and literary scholars, among others, share an interest with information scientists, as May Yuan demonstrates in her exploration of how to spatialize text in geospatial tools. The semantic web, so necessary for artificial intelligence, in its spatial form has much to contribute to deep mapping as well. Integrating this spatial knowledge occurs in three ways, Nicholas Bauch reminds us: it brings together different formats, links entities, and combines abstract and embodied geographic knowledge about a place. This creative integration, he contends, provides opportunities for encounters that engage the senses and the intellect.

Representing this integration effectively poses a challenge, however. Charles Travis is loath to abandon GIS, which indeed is a powerful software, but is keen to enhance it by drawing on older cartographic traditions of poetic geography and geosophy. He applies the insights from these traditions to a historical event, the Irish Rebellion of 1916, with special attention to its visualization. Patricia Murrieta-Flores, Mariana Favila-Vázquez, and Aban Flores-Morán remind us that deep mapping is well suited to understanding the world as socially constructed, in their instance by comparing indigenous and western cartography as evidenced in sixteenth-century Mesoamerica. Trevor M. Harris presents a different version of deep mapping, one that embraces more fully the phenomenology of space and place. Drawing on recent advances in technology, he explores the role of interactive virtual platforms that immerse the virtual umwelt, a sensuous 3D environment capable of recreating elements of the spatially complex lived world, and that provide agency to the reader to explore, examine, understand, and gain insight from these representations.

But for what purpose do we pursue deep maps? Stephen Robertson and Lincoln A. Mullen suggest that the ability to trace a narrative path within and among the multiplicity of voices embedded in a deep map takes us to the heart of what humanists want to address, the question of why. Various tools exist to support this narrative thrust, and they explore a project, Digital Harlem, that helps us understand how deep mapping supports our need to tell a multifaceted story. Mia Ridge, a computer scientist who works at the British Library, finds another role for deep maps: they provide both infrastructure and contextual interface for the ever-growing digital collections of cultural heritage institutions. Deep maps, it seems, not only embrace multiplicity, they also have multiple uses. John Corrigan explores deep mapping as an approach to research. This form, he argues, is a "fluid assemblage" that mirrors the critical discourse of the humanities and both complements and advances scholarship. Finally, Karen Kemp provides a personal reflection on the development of deep mapping and its relationship to GIScience.

These essays provide guidance and examples for readers interested in deep mapping as practice and product, but they do not exhaust the subject. Even a cursory glance will discover a rapidly expanding literature that is redefining the potential for this approach as well as offering testbeds in the forms of projects that will shape the way we think about deep mapping. This volume offers a gateway to these developments and a way to think more productively about both the how and why of deep mapping.

A note regarding referencing conventions used in this book. With authors from various disciplines, we have chosen not to impose a common citation scheme but have allowed the use of whatever conventions are common to their field. In this way we reduce both the chance of error and the time required to ensure a consistency that no longer seems useful.

Collaborative works do not appear simply by the good graces of the participants. They require much planning and considerable financial support. *Making Deep Maps* stemmed from a workshop held in May 2016 at West Virginia University and supported generously by a grant from West Virginia University. The authors gratefully acknowledge this funding as well as the gracious assistance of Barbara McLennan, Frank Lafone, and Deborah Kirk in ensuring that the workshop ran smoothly.

The Editors
September 2020

Notes

1 See David J. Bodenhamer, John Corrigan, and Trevor M. Harris, eds., *Deep Maps and Spatial Narratives*. Series on Spatial Humanities (Bloomington, IN: Indiana University Press, 2015).
2 For more on feminist geography, see Linda McDowell, *Gender, Identity, and Place: Understanding Feminist Geographies* (Cambridge, UK: Polity Press, 1999) and Lisa Nelson and Joni Seager, eds., *A Companion to Feminist Geography* (Malden, MA: Blackwell Publishing, 2005).

1 The varieties of deep maps

David J. Bodenhamer

In most accounts, deep mapping is a recent development with a straightforward genealogy. It begins with a call from Situationists, avant-garde French intellectuals in the 1950s, for an affective geography that combined material and emotional features of a place in a contextual stew. Blending unitary urbanism and psychogeography, they aimed to "record and represent the grain and patina of place through juxtapositions and interpenetrations of the historical and the contemporary, the political and the poetic, the discursive and the sensual."[1] Four decades later, American author William Least Heat-Moon gave shape to this concept in *PrairyErth*, a stratigraphic pastiche of a small Kansas county. His subtitle, (*a deep map*), also gave the concept a name, and the question he asked—"Would coordinates lead to connections?"[2]—announced its purpose. The term deep mapping thus entered our vocabulary and stirred our imagination. A new field, spatial humanities, and advanced geospatial technologies gave this approach both momentum and a new expression, with an increasingly rich literature describing both its potential and challenges.[3]

If only it were that simple or that recent. The impulse toward deep mapping—linking geographic and cultural representations of a place—began when mankind first created maps. One of the earliest known examples, the *Çatalhöyük Map* from Turkey (ca. 6200 BCE), displays an arrangement of differently sized dwellings, possibly representing social status, and a far out-of-position volcano that had both religious and economic significance for the inhabitants.[4] Some leading cartographers dispute its label as a map because it is part of a wall painting that depicts other features of communal life,[5] but even if so, it demonstrates that people did not separate their physical and cultural worlds. The ancient Greek geographer Ptolemy labeled this practice chorography, the detailed description of a location, which reflected "the qualities rather than the quantities of the things that it sets down."[6]

This urge to provide contextual information was common in medieval maps with their elaborate cartouches, imaginative renderings, and embedded texts. Consider the thirteenth-century Ebstorf Map, the largest example from a genre known as *mappae mundi*. The map, an idealized depiction of the known world sewn from thirty goatskins, is a visual encyclopedia of knowledge about the world, with over 2,000 drawings and explanatory texts that combine geography, religion, science, and folklore.[7] In it the earth is an event-filled and storied place. It was a deep map.

DOI: 10.4324/9780367743840-1

The revival of chorography in the Renaissance, especially in the form of detailed written descriptions of places, also reflected this impulse.[8]

The rise of the nation-state, with its territorial ambition, and more recently the industrial-bureaucratic state, fostered the development of the map that we recognize as modern. It was spare, depicting routes and boundaries, as well as physical features, which facilitated or impeded the movement of people and goods. Although these flat maps became ubiquitous—in fact, they recast the popular meaning of mapping itself—the need for information about place did not diminish; rather, it took other forms.

The development of these forms reveals how widespread are the roots of deep mapping. Eighteenth-century gazetteers and antiquarian local histories were a smorgasbord of history, folklore, statistics, natural history, and anything else that could be discovered about a place. Even when they did not contain maps, gazetteers embodied the impulse toward contextualization that has marked thinking about deep maps. City encyclopedias that flourished at the end of the nineteenth and twentieth centuries were variants on this theme. So too was the *Annales* school of historiography that insisted on a thick description of the past that included the physical and material culture in which actions occurred. Intellectual influences extend to Martin Heidegger's insight that we naturally use the body as a reference point to organize space[9] and Michel de Certeau's spatial stories that captured the practices of everyday life.[10] Theorist Edward Casey insisted that we are continually making place by being continually in a "configurative complex of things,"[11] and geographer Yi Fu Tuan emphasized the connection between emotion and landscape seen in the love of place (topophilia).[12] Doreen Massey contributed a notion of space as open, hybrid, and marked by interconnecting flows,[13] and Michael Pearson's and Michael Shanks' practice of theatre/archaeology refashioned fragments of the past as real-time events.[14] We have been developing deep mapping, it seems, for a long time.

Postmodernism, the spatial turn, and deep mapping

Two of the most influential intellectual trends of the past half century, Postmodernism and Deconstructionism, have provided impetus for the interest in deep mapping. Their advocates claimed that rational empiricism were societal artefacts, with its truth merely one discourse among many. Deconstructionists especially insisted that language constructs reality; people who use language differently in fact see the world differently. This social construction of reality marked a "paradigmatic shift from the Enlightenment belief in the relative indeterminacy of both the natural world and the social world to the—increasingly widespread—post-Enlightenment belief in the radical indeterminacy of all material and symbolic forms of existence."[15] One result has been a scholarly insistence on recognizing—and reflecting—a fuller awareness of the rich diversity of human experience.

The spatial turn is an expression of this postmodern world. The phrase has a murky lineage, but it has become common shorthand, especially in the social sciences, where it refers to a revival of interest in the relationship among space, culture, and social organization. As part of a postmodern shift that rejected the universal truths, grand narratives, and structural explanations that dominated twentieth-century social sciences and humanities, the spatial turn was about the particular and local, without any supposition that one form of culture is better than another. Its claim was straightforward: to understand human society and culture we must understand how it developed in certain circumstances and in certain times and at certain places.

The reintroduction to space began in the 1970s and 1980s, when a new critical geography emerged. "Rather than being seen only as a physical backdrop, container, or stage to human life," Edward Soja, a leading theorist noted, "space is more insightfully viewed as a complex social formation, part of a dynamic process."[16] This sense of space as social process and not simply geography owed much to postmodernism, which viewed the world as embodying extreme complexity, contradiction, ambiguity, uncertainty, and diversity. These ideas were not new—the nineteenth-century German philosopher Friedrich Nietzsche famously noted that "whichever interpretation prevails at a given time is a function of power and not truth"—but postmodernism gave them a different expression.[17] For postmodernists, the way we see and define the world is unavoidably relative; in spatial terms, this stance meant that every society defined space differently according to its needs. Power and culture influenced the way societies thought about space, just as the potential and limits of the material world shaped the way people constructed their societies. Feminist geography was especially important in advancing conversations about the urgency of attending to lived experience, complexity, and post-positivist discourses of society and knowledge.[18]

The postmodern spatial turn also was marked by a critique that geography had been devalued in favor of historical experience, or time, in twentieth-century scholarship: "Space was treated as the dead, the fixed, the undialectical, the immobile," argued Michel Foucault, "[whereas] Time . . . was richness, fecundity, life, dialectic."[19] By late century, space as a new interpretive framework had found its way back into social theory, or as David Harvey has noted, "geography [was] too important to be left to geographers."[20] Now, space was viewed as the medium for the development of culture, containing embedded stories based on what has happened there. These stories are both individual and collective, and each of them link geography (space) and history (time). More important, they reflect the values and cultural codes present in the various political and social arrangements that provide structure to society. Space in this broader meaning became a primary way to understand changes such as globalization, war, and cultural conflict. As Barney Warf has noted, geography is not an afterthought to history; events are bound to their spatial context.[21]

New spatial technologies, especially Geographic Information Systems (GIS), facilitated a (re)discovery of geographical space in the study of the past. Yet for all its capabilities, skeptics maintained, GIS offered a view of the physical environment stripped of its cultural assumptions. It allows us to know where something occurs and to see what else is happening in the same space, but it tells us nothing about the meaning of what we see. It also delineates space as a set of coordinates with attached attributes, a cartographic concept, rather than as relationships, a social concept. These tendencies excluded non-Western conceptions of the world. American Indians, for example, defined the world as a set of interlinked phenomena, only some of which can be defined as geographic space.[22] It is easier to understand ancient China dynasties when we accept their definition of space as networks of places and actors rather than as prescribed jurisdictions with formal boundaries.[23] GIS had difficulty managing these different meanings of space. It was, at heart, a tool for quantitative data.

These epistemological and ontological differences led to an uneasy rapprochement in the early twenty-first century about how to represent space. The demands of a modern industrial and technological society required fixed space, a Euclidian concept.[24] This literal representation of the world commodified cartography and benefitted consumers who relied on location-based services from Google Maps to Yelp, but statistical mapping was at odds with an Einsteinian view of relative space–time. As Henri Lefebvre observed, we "live in a world of Euclidian and Newtonian space while knowledge moved in the space of relativity."[25] Computer cartography portrayed the world as geo-coded, whereas both geographic science and the humanities increasingly viewed the map as a dynamic representation or interpretable text. The problem was that GIS did not allow scholars to see this dynamic world easily, if at all.

The notion of dynamic and relative space challenged cartographic conventions, and deep mapping emerged as a response to this problem. Deep maps are visual, multilayered, and inclusive, offering multiple perspectives of a small area of the earth. They are inherently unstable, continually unfolding and changing in response to new data, new perspectives, and new insights. Open to experience and not simply to measurement, they are framed as a conversation and not a statement. They are above all dynamic constructions, not only revealing the fluid and complex nature of human actions that shaped and are shaped by space but also allowing us to see the deep contingencies and particularities of place.[26]

Technological convergence and deep mapping

Traditional GIS with its emphasis on precise measurement and spatial models is not the means for constructing a deep map. But it is part of the solution, especially when linked to recent advances in spatial multimedia, GIS-enabled web services, cyber geography, and virtual reality, among other tools. Geo-visualization offers an approach useful to humanists because its arena is mental representations, not maps of precisely measured objects; it also aims not to chart only what is known but to explore as well what is unknown.[27] Collectively, these technologies allow us to probe the situated knowledge that resides in dynamic and contested memories and to understand the affective or emotional geographies of space and place.[28] They have the potential, in brief, to revolutionize the role of place in the humanities by moving beyond the two-dimensional map to explore dynamic representations and interactive systems that will prompt an experiential, as well as rational, knowledge base.

The coalescence of digital technologies over the past decade makes it possible to envision how a mix of geospatial and immersive technologies contributes to the formation of a deep map. Archaeologists have used GIS and computer animations to reconstruct the Roman Forum, for example, creating a 3D world that allows users to walk through buildings that no longer exist, except as ruins. We can experience these spaces at various times of the day and seasons of the year. We see more clearly a structure's mass and how it clustered with other forms to mold a dense urban space. In this virtual environment we gain an immediate, intuitive feel for proximity and power. This constructed memory of a lost space helps us recapture a sense of place that informs and enriches our understanding of lost places, such as ancient Rome (*Digital Roman Forum Project*). In similar fashion, historians and material culturists have joined with archeologists to fashion *Virtual Jamestown*.

In the last decade, the development of the geospatial web has made mapping available and accessible to digital historians, with web mash-ups allowing maps to be created and iterated with unprecedented ease. Web mapping platforms such as Google Maps, MapQuest, CartoDB, Palladio, and Neatline do not have the same quantitative orientation as GIS software, allowing a diversity of spatial data to be visualized. More advanced and customized web mapping is also now possible using open-source tools such as Leaflet, GeoServer, d3.js, and OpenLayers.[29] With these tools, historians explored a range of questions and topics: urban development, housing, ghettos, moviegoing, language use, sound, industries ranging from iron to cutlery, events such as bombing, military recruitment, the slave trade, slave revolts and manumission, the spread of disease, witchcraft accusations, feminist activism, and the circulation of correspondence.[30] Although not deep maps in the fullest sense of the term, these efforts are pushing past points and polygons to construct a complex spatial past. The result is "thick mapping," a term advanced by creators of *HyperCities*, an innovative GIS-based website that allows scholars to explore cities across time, using period maps overlaid on contemporary landscapes, viewing virtual buildings within their real-world context, and linking images and text to the locations they describe.[31]

The commercial market also is rapidly developing tools that can be appropriated for deep mapping. Consider Oculus Rift, the virtual reality technology owned by Facebook, that permits users to have a distortion-free, immersive view of an environment enhanced by information and images harvested from the Web. Gaming engines also offer ways to reconceptualize the role of space in the humanities by privileging agent-based exploration rather than linear movement as a means of discovery. Spatial stories in this environment are "held together by broadly defined goals and conflicts and are pushed forward by the character's movement," not by the structure of an argument.[32] Truth and authenticity are measured not by standards of causality but by the game's ability to create experiences of space that expand and improve our understanding of a complex and multifaceted reality. This convergence of technologies invites us to move far beyond the static map, to shift from two dimensions to multidimensional representations, to develop interactive systems, and to explore space and place dynamically—in effect, to create virtual worlds embodying what we know about space and place.

Definitions and typologies

What these developments suggest is that we are at the point in our conceptualization of deep mapping to think more critically about definitions and typologies. Like the spatial humanities, its parent discipline, deep mapping fuses the humanities' traditional focus on nuance, voice, experience, text, and image with the systematic approaches of spatial science, computer modeling, and virtual reality. Its aim is to link time, space, and culture dynamically.

This goal is not new, nor is it uniquely postmodern. It addresses the problem noted almost a century ago by Mikhail Bakhtin, Martin Heidegger, Edmund Husserl, and other theorists who recognized the disjuncture between the Enlightenment world of logical–rational empiricism and the nascent world of Einsteinian relativity and phenomenology. Writing in the 1920s and 1930s, Bakhtin, a Russian semiotician and literary theorist, captured the problem neatly in his notion of the chronotope, the inextricable linkage of time and space that defines both reality and our perception of it. As Bakhtin describes it, the chronotope is an experiential form:

> [T]ime, as it were, thickens, takes on flesh, becomes artistically visible; likewise, space becomes charged and responsive to the movements of time, plot and history. . . . [It] . . . is the place where the knots of narrative are tied and untied. . . . Time becomes, in effect, palpable and visible; the chronotope makes narrative events concrete, makes them take on flesh, causes blood to flow in their veins.[33]

For Bakhtin, the world could not be understood without considering both when and where an event or action occurred, as well as when and where observers stood as they became aware of or considered an event. By its very nature, the time–space link within the chronotope emphasizes particularity; it reminds us of the distinction between the observed and the observer; and it reflects the dynamic, interdependent, and relative context of all knowledge. Although he did not reference Bakhtin, literary theorist Franco Moretti was reaching for something similar when he called for maps that are:

> more than the sum of their parts: they will possess "emerging" qualities, which were not visible at the lower level. . . . Not that the map is itself an explanation, of course: but at least, if offers a model of the narrative universe which rearranges its components in a non-trivial way, and may bring some hidden patterns to the surface.[34]

Deep mapping allows us to corral Bakhtin's ghost and realize Moretti's notion of a map that reflects emerging realities. It invites a phenomenological conception of place deeply invested in human experience.[35]

How then do we define a deep map? Perhaps the simplest definition is this one: Deep maps are fluid cartographic representations that reveal the complex, contingent, and dynamic context of events within and across time and space. They express the results of research but the answers they provide are not final and the map itself is always open to revision. Yet regardless of their form, deep maps have certain distinguishing characteristics: they link time and space (chronotope), operate across multiple scales of time and space, embody multiple agents and multiple perspectives, recognize alternate schemes and emergent realities, foster a dynamic context that reveals movement and linkage, and (ideally) are emotional and experiential. Although other cartographic forms may reflect one of more of these features, deep maps exemplify most or all of them.

At its core, a deep map is a hybrid construction, and although it relies on new technologies, it is not primarily a computational product. Rather, it enlists the approaches and methods of humanists—memory, narration, curation, and knowledge design, among others—as much it does ontologies, coding, or 3D modeling. It aims to create a densely layered, thickly embedded, and highly contingent environment that allows us to explore the development of place in all its spatiotemporal complexity, but we achieve this end in multiple ways. But regardless of form, deep maps encompass five core principles: they are *flexible*, inviting exploration; *user-centric*, supporting differing views; *path-traceable*, supporting narration; *open*, admitting new material; and *immersive*, evoking experience.

In addition to these characteristics, a deep map also can be any one of three things—a platform, a process, and a product—or all three at once. As platform, it is an environment embedded with tools to bring data into an explicit and direct relationship with space and time. As a process, it engages evidence within its spatiotemporal context and traces paths of discovery that lead to a spatial narrative and ultimately a spatial argument. As product, it is the way we make visual the results of our enquiry and share the spatially contingent argument enabled by the deep map. Within this environment, we can develop the event streams that permit us to see the confluence of actions and evidence; we can use path markers or version trackers to allow us (and others) to trace our explorations; and we can contribute new information that strengthens or subverts our argument, which is the goal of any exploration.[36] A deep map, in short, is a new curated and creative space.

Yet not all deep maps are alike. Instead, the deep map will be responsive in its form to the questions asked, the evidence gathered, and the limits of the technologies used to create it. But it is possible to identify at least five categories, or types, of deep maps—or, perhaps, deeper maps—and to offer examples as illustration.[37] In ascending order of complexity, these types are:

1. *Archival*: spatially and temporally enabled open repository that allows access to quantitative, qualitative, and image data. This category defines the most basic deep map; it includes some but not all the characteristics of the genre. Often based on a GIS, but not always, it includes both base maps and historical maps, some of which may already be geo-rectified. Spatially enabled data can then be mapped on these layers, which can be combined in ways that are standard to GIS. These data can be quantitative, qualitative, and images, each with spatiotemporal footprints of varying precision.

Many national historical GIS projects fit this category. The German Historical GIS (HGIS Germany), for example, not only provides a mapped view of the continually changing boundaries of the multiple political entities in nineteenth-century Germany but also allows users to identify the ever-changing dynastic alliances, to add demographic and economic data, and to view available multimedia files.[38] By the early twenty-first century, the United States, Germany, Netherlands, Belgium, Russia, South Korea, and China (one from the People's Republic and one from Taiwan), among others, boasted a national historical GIS.[39] Elsewhere, variants of this genre emerged, usually based on particular themes, for example, The Cultural Atlas of Australia, or on cities, with Sydney, London, and Tokyo among the best examples.[40]

Not all examples in this category use GIS or have a map-based interface. The Digital Public Library of America (DPLA), a repository of texts, images, maps, and other materials from contributing libraries and archives across the United States allows users to access material by time, space, and subject. More important are the apps that accompany the site, as well as the open APIs that allow developers to create new ways of manipulating DPLA content, including linking its content with that of other major repositories such as Europeana. These apps facilitate the curation that is part of any deep map. Google's Arts and Culture platform goes a step further by allowing the user to create a thematic presentation of this curated information.[41]

A public-facing archival deep map is RICHES, an umbrella program at University of Central Florida that brings together academic public history projects with community-based resources through a federated structure that links digital collections in Central Florida. It provides a searchable database with access to images, documents, podcasts, oral histories, films, and visualizations. Significantly, it combines time and space with text analysis techniques to find hidden connections within the archive. It also allows users to contribute their material (e.g., letters, diaries, photographs, etc.) to the federated archives, as well as to create exhibits that link time and space.[42]

2. *Descriptive*: densely layered map of a place with information drawn from multiple sources; designed to describe context or theme. Here the emphasis is on a curated product that seeks to advance a more complex view of place. The sites often use GIS, although not exclusively, and frequently include elements of an archival deep map.

An early example was the *Valley of the Shadow*, a digital history project that created dense views of two Shenandoah Valley counties, one in slave-holding Virginia and the other in free-state Pennsylvania, during the period immediately preceding and during the Civil War. In addition to maps of terrain, infrastructure, and battles, many of them augmented with additional information; the map of roads, for example, also notes residences along these routes. Other parts of the site include letters and diaries, newspapers, military records, and the other archival data. No longer an active site, the project allowed users access to rich descriptive material from which they could construct arguments. The *Salem Witch Trials* site, also a documentary and transcription archive, makes a more complete nod toward mapping by providing spatio-temporal explorations of the relationship between accused and accusers, traceable through accompanying documents. The same is true for virtual site reconstructions, typically from archaeology, exemplified by *Mapping Medieval Chester*, which provides dynamic, diachronic views of the city from 1200 to 1500, focusing on archaeology, built space, and literature in this historic border town.[43]

The Great Britain Historical GIS, an archival deep map like other HGIS efforts (even though it was the first such site), found public expression in *Vision of Britain*, a companion website that allowed a spatial exploration of descriptive information about British places.[44] It allows users to access material on over 23,000 British places through a variety of lenses—maps, statistics, reports, travel accounts, and images, among other types of data—from 1801 to the present. In some ways it is the digital equivalent of eighteenth-century atlases, including a place–name gazetteer, but the site is more than a storehouse of information; it also allows the user to trace change over time for the British places of interest. Above all, it provides a thick description of places—a thick map, to use Todd Presner's term[45]—that uses GIS only in part.

3. *Narrative*: structured view of spatially and temporally located data; designed to facilitate or make a scholarly argument within a rich spatio-temporal, mapped environment.

Humanists traditionally advance knowledge about place through narrative, which encourages the interweaving of evidentiary threads and to use emphasis, nuance, and other literary devices to achieve the complex construction of past and present worlds. We are continually making place by our acts of living—geographer Doreen Massey argues that it is the result of interconnecting flows and not something that is rooted or fixed—but written narrative is linear, not recursive.

The narrative deep map allows us to mimic or at least suggest the fluid complexity of place, as Nicholas Bauch (who has a chapter in this volume) demonstrated in his pioneering e-book, *Enchanting the Desert*.[46] Bauch took an early twentieth-century series of iconic photographs of the Grand Canyon and asked how these images had reshaped the cultural history of the place, in effect remaking space in a European American form. Using GIS viewsheds, he reveals the silences in the images, the spaces hidden from view, and allows readers to view the photographs, viewsheds, and their cultural history at once, with the ability to bring various facets of the narrative to the foreground depending on what questions are asked. What emerges is a critical cartography with a palimpsest of meanings that have accrued over nearly 10,000 years. Space and place emerge as the product of inter-relationships, coexistence, and process, always in the state of becoming, with the reader invited to enter the narrative and test its argument.

4. *Exploratory*: dynamically constructed, path-traceable maps of places and place-based themes. Much closer to an ideal type, these deep maps are more flexible and open-ended than archival and descriptive sites. They too use GIS but often in concert with graph databases or linked open data models that open them to information added by members of a research community.

A web-based collaboration among several Dutch universities and archival services, *Circulation of Knowledge and Learned Practices in the 17th Century Dutch Republic*,[47] links correspondents by place and time in an epistolary network to understand how knowledge spread in early modern Holland. The search tools lead to interactive maps of senders' and recipients' locations, times, and correspondent and corecipient networks. Similarly, the *Mapping the Republic of Letters* used mapped, dynamic networks to trace the spread of knowledge in the European Enlightenment.[48] Neither project uses GIS, although both are spatial and temporal in their approach.

A linked open data model allows *Pelagios Commons* to create an exploratory deep map that brings together a variety of place-based, thematic web resources.[49] HTML has always allowed links but only in one direction; linked open data creates multi-lateral connections among a variety of historical content—textual, image, maps, and the like—related to a specific concept. Equally important, it is decentralized, that is, the linked data expands and is repurposed through the contributions of users. The commons, a user-centric portal, provides access to tools to find, record, and export data, as well as map tiles that allow these data to be displayed dynamically. Although the immediate focus is on the ancient world, the approach can be extended to other subjects. By annotating web resources to include geographic information, users can link and share data effectively, thereby curating new perspectives on themes of interest to the community.

5. *Immersive/experiential*: focus on presence, a sense of reality, with its contextual awareness of past actions and embedded contingencies; seeks to place user within a spatially and temporally sensitive environment as a way of reifying context; fosters an aesthetic and emotional experience with place.

Trying to comprehend space, place, and time in concert has always proven difficult, even in the most expert narratives. Historian Hugh Trevor-Roper noted the problem decades ago:

> How can one both move and carry along with one the fermenting depths which are also, at every point, influenced by the pressure of events around them? And how can one possibly do this so that the result is readable?[50]

It is the same problem that prompted Bakhtin's conception of the chronotope. Technology now makes it possible to create an environment that approximates Bakhtin's ideal, a deep map allowing users to experience the story even as they seek to construct a story of experience. It goes beyond attempts to capture the movement of people through space, which was at the heart of geographer Torsten Hägerstrand's notion of time geography with its representation as space–time cubes, but rather to recognize that we live in a multidimensional world that includes memory and expectation—past time and future time—and not simply observed events. How then do we represent this world—and what does it add to our understanding of human actions, behavior, and motivations?

Augmented and virtual reality are two approaches to the immersive deep map. Major companies (e.g., Facebook and Google) are betting on the role of these technologies to serve commercial ends, so it is likely that cost and complexity will lessen as barriers to their use in scholarship as well. Archaeologists have experimented with AR for well over a decade as a way of enhancing their models through real-world experience.[51]

More ambitious efforts involve virtual reality. To aid in trials of Nazi war crimes, the Bavarian state criminal prosecutor's office has recreated the Auschwitz-Birkenau concentration camp in a highly detailed VR to help judges and juries understand more fully what happened there.[52] Bema is a multimodal user interface that enables scholars of Greek rhetoric and oratory to perform virtual reality studies of ancient political assemblies at the hill of the Pnyx by gaining a real-world sense of what speakers could see and listeners could hear, as well as how structural changes in the amphitheater over time influenced behavior.[53] *Virtual Plasencia* uses a gaming engine to study the interactions of three religious traditions—Islam, Christianity, and Judaism—in fifteenth-century Spain.[54] At West Virginia University, Trevor M. Harris—see his chapter in this volume—has moved GIS into the CAVE (Computer Augmented Virtual Reality) and recreated nineteenth-century Wheeling, West Virginia. Here, more senses are available for understanding place: not only does the user have the physical sensation of walking through the town, created in Google Sketch-Up, but Harris and his students have also introduced visual effects, such as smoke-generated haze, and the smells of the coal-burning furnaces and abattoirs. These effects were not mere tricks to delight the user. Rather they helped readers grasp more fully what Rebecca Harding Davis, a nineteenth-century American novelist, intended in her evocative *Life in the Iron Mills* (1861), when she asked, "do you know what it is like in a town of iron works? . . . [C]ome right done with me, here, into the thickest of fog and mud and foul effluvia."[55]

The experience of modern gamers suggests the rich potential of immersive deep mapping. *EVE Online* is a futurist game in which users create avatars from one of four races, choosing hair, scars, and clothes to distinguish individuals. The game revolves around three basic tasks: mining asteroids for minerals, using these minerals to build spaceships, deploying these vessels in battle. It is a dystopian world made up of thousands of interlocking solar systems, with independent agents to monitor safe zones but with vast spaces in which anything goes. It also has an economy but no legal system; everything depends upon trust and relationships. Now, many games provide similar experiences but *EVE Online*, an unbounded (no rules) massively multiplayer gaming environment, has been studied extensively by scholars. The findings are interesting—and suggestive: half a million players inhabit this universe at the same time and they continually make both the rules and interpret the meaning of their experiences. Users internalize the history of this fictive universe; they keep records of events, interview famous players, and write their own narratives of what has happened. They are immersed in *EVE* and make sense of it through storytelling, with the multiple narratives allowing us to see the world from multiple perspectives. There is no master argument, no voice of authority.[56] Instead, like the deep map, its multivocality leads to negotiated meanings that reveal the contingent nature of this fictive yet all-too-real world.

The value of deep maps

Regardless of the type, deep maps will depend upon scholarly traditions developed over generations by the various humanities disciplines. They must reflect the ways we discover knowledge even as they contribute new approaches to research questions. A deep map will not replace close reading of a text, for example; rather, deep mapping exists alongside other methods of investigation, whether text mining or literary cartography or the text as artefact, as we seek to gain a more complete purchase on our inquiry. It also welcomes knowledge developed in nontraditional ways, such as reflected in neogeography, the nonexpert or volunteered spatial information produced in everyday life, which may not follow the vetting practices or publishing formats normally associated with our disciplines.[57] Scholarship may take new, more fluid forms but the aims of the monograph and essay will not be replaced or subsumed as much as they will be reimagined and reexpressed.

What the deep map allows—no, encourages—is what John Corrigan (also in this volume) elsewhere has termed a "genealogy of emplacement." That is, it frames investigations of space and place as "a kind of genealogy . . . into the historically contingent layering of structures, strategies, tactics, discipline, anti-discipline, environment, everyday practice."[58] This genealogy deemphasizes the predictably rational by focusing instead on locally specific and unique processes, as well as in interstices between everyday practice and structure. Although it embraces some of William James's sense that "reality, life, experience, concreteness, immediacy . . . exceeds our logic, overflows, and surrounds it,"[59] its aim is to understand how place is made through everyday practice and how it influences (and is influenced by) structure. In its observations of practice, emplacement also locates the networks and associations that constitute narratives of place. The deep map, in this sense, narrates through curation, constructing complex collages that reveal change over time as well as the inherently varied meanings we ascribe to place. Its means of representation is the spatial story that emerges from this act of curation.

A deep map, in brief, promises a different way to understand society and culture, past and present, fictive or real. Historian William Sewell, for instance, has argued that "social life may be conceptualised as being composed of countless happenings or encounters in which persons and groups of persons engage in social action. Their actions are constrained and enabled by the constitutive structures of their societies."[60] As a result, "'societies' or 'social formations' or 'social systems' are continually shaped and reshaped by the creativity and stubbornness of their human creators."[61] The deep map embodies these human engagements and the structures that accommodate (or flow from) them. Deep maps have the potential to help us see the interactions of agents and structures in the narratives we construct, all bound in the space–time that defines place.

Will deep mapping enhance scholarship? It is a legitimate question. Making the process experiential and open-ended may lead to greater understanding; it also may provide so many narratives, so many voices, that it fragments our understanding and results in intellectual chaos. It may add nothing more than our essays and books provide collectively, except perhaps for whatever utility is gained through interactive, scalable visualizations of a wide array of evidence. But what the deep map offers far outweighs the risk. At its best, a deep map opens discovery and invites multiple expert and naïve voices into the conversation about meaning. It reinforces the role of emotion and memory in the construction of place and event, not to the exclusion of rational argument but as a different way of understanding. It allows us to uncover the emergent realities and deep contingencies of the past that arise from the intentional and unintentional actions of people as they make their way through life and create the stories that define us. In the process, it enables a unique postmodern scholarship that is rarely possible, if at all, through traditional disciplinary forms.

Notes

1 Mike Pearson and Michael Shanks, *Theatre/Archaeology* (London: Routledge, 2001), 64–5. Also see David J. Bodenhamer, "Creating a Landscape of Memory," *International Journal of Humanities and Arts Computing* 1, no. 2 (2008): 97–110.
2 William Least Heat-Moon, *PrairyErth (A Deep Map)* (Boston: Houghton Mifflin Company, 1991), 15.
3 See, for instance, David J. Bodenhamer, John Corrigan, and Trevor M. Harris, eds., *Deep Maps and Spatial Narratives* (Bloomington, IN: Indiana University Press, 2015).
4 S. Meece, "A Bird's Eye View: Of a Leopard's Spots: The Çatalhöyük 'Map' and the Development of Cartographic Representation in Prehistory," 2005. www.dspace.cam. ac.uk/handle/1810/195777.
5 John Krygier, "Cartocacoethes: Why the World's Oldest Map Isn't a Map," 2008. https:// makingmaps.net/2008/10/13/cartocacoethes-why-the-worlds-oldest-map-isnt-a-map/.
6 J.L. Berggren and Alexander Jones, eds., *Ptolemy's Geography* (Princeton: Princeton University Press, 2000), 57–8.
7 Hartmut Kugler, ed., *The Ebstorfer Weltkarte*, 2 vols. (Munich: Oldenbourg Academy, 2006).
8 Lucia Nuti, "Mapping Places: Chorography and Vision in the Renaissance," in Denis Cosgrove, ed., *Mappings* (London: Reaktion Books, 1999), 90–108.
9 Martin Heidegger, *Being and Time: A Translation of Sein und Zeit*, trans. Joan Stambaugh (New York: State University of New York Press, 1996), 97.
10 Michel de Certeau, *The Practice of Everyday Life*, trans. Stephen Rendall (Berkeley: University of California Press, 1984).
11 Edward S. Casey, "How to Get from Space to Place in a Fairly Short Stretch of Time: Phenomenological Prolegomena," in Steven Feld and Kenneth H. Basso, eds., *Senses of Place* (Santa Fe, N.M.: School of American Research Press, 1996), 25.
12 Yi Fu Tuan, *Space and Place: The Perspective of Experience* (Minneapolis: University of Minnesota Press, 1977).

13 Doreen Massey, "A Global Sense of Place," in Trevor Barnes and Derek Gregory, eds., *Reading Human Geography* (London: Hodder Arnold, 1997), 315–23.

14 Mike Pearson and Michael Shanks, *Theatre/Archaeology* (Abingdon, UK: Routledge, 2001).

15 Simon Susen, *The 'Postmodern Turn' in the Social Sciences* (New York: Palgrave Macmillan, 2015), 1.

16 Edward Soja, "In Different Spaces: Interpreting the Spatial Organization of Societies," Proceedings 3rd International Space Syntax Symposium, Atlanta, 2001.

17 Postmodernism's progenitors go well back to the first part of the twentieth century, for example, Dadaist efforts to destroy the categories of high and low culture, Heidegger's rejection of notions of objectivity subjectivity, and existentialism's doubt, ambiguity, and uncertainty, among many others. For a brief critical overview, see Christopher Butler, *Postmodernism: A Very Short Introduction* (Oxford: Oxford University Press, 2003).

18 See Lisa Nelson and Joni Seager, eds., *A Companion to Feminist Geography* (Malden, MA: Blackwell Publishing, 2005).

19 Michel Foucault, *Power/Knowledge: Selected Interviews & Other Writings, 1972–1977*, ed. Colin Gordon (New York: Pantheon Books, 1980), 70.

20 David Harvey, *Spaces as Capital: Towards a Critical Geography* (Abingdon, UK: Routledge, 2001), 116.

21 Barney Warf and Santa Arias, "Introduction: The Reinsertion of Space in the Humanities and Social Sciences," in Barney Warf and Santa Arias, eds., *The Spatial Turn: Interdisciplinary Perspectives* (Abingdon, UK: Routledge, 2009), 1–10.

22 R.A. Rundstrom, "GIS, Indigenous Peoples, and Epistemological Diversity," *Cartography and Geographic Information Systems* 22 (1995): 45–57.

23 Merrick Berman, "Boundaries or Networks in Historical GIS: Concepts of Measuring Space and Administrative Boundaries in Chinese History," *Historical Geography* (2005): 118–33.

24 Denis Cosgrove, "Maps, Mapping, Modernity: Art and Cartography in the Twentieth Century," *Imago Mundi: International Journal of Cartography* 57 (2005): 35–54.

25 Henri Lefebrve, *Critique of Everyday Life, vol. III: From Modernity to Modernism (towards a Metaphilosophy of Daily Life)* (London: Verso, 2005), 46.

26 David J. Bodenhamer, John Corrigan, and Trevor M. Harris, "Deep Maps and the Spatial Humanities," in David J. Bodenhamer, John Corrigan, and Trevor M. Harris, eds., *Deep Maps and Spatial Narratives*. Series on the Spatial Humanities (Bloomington, IN: Indiana University Press, 2015), 1–5.

27 Alan M. MacEachren, Mark Gahegan, and William Pike, "Visualization for Constructing and Sharing Geo-Scientific Concepts," *Proceedings of the National Academy of Sciences* 101 (2004): 5279–86.

28 James Craine and Stuart Aitken, "The Emotional Life of Maps and Other Visual Geographies," in Martin Dodge, Rob Kitchin, and Chris Perkins, eds., *Rethinking Maps: New Frontiers in Cartographic Theory* (New York: Routledge, 2009), 168–85.

29 Stephen Robertson, "The Differences between Digital Humanities and Digital History." http://dhdebates.gc.cuny.edu/debates/text/76.

30 "DH GIS Projects." http://anterotesis.com/wordpress/mapping-resources/dh-gis-projects/.

31 Todd Pressner, David Shepard, and Yoh Kawano, *HyperCities: Thick Mapping in the Digital Humanities* (Cambridge, MA: Harvard University Press, 2014).

32 Henry Jenkins, "Game Design as Narrative Architecture." http://web.mit.edu/cms/People/henry3/games&narrative.html.

33 M. Bakhtin, *The Dialogic Imagination: Four Essays*, ed. M. Holquist (Austin, TX: University of Texas Press, 1981), 250.

34 Franco Moretti, *Graphs, Maps, Trees: Abstract Models for a Literary History* (London: Verso, 2005), 53–4.

35 Eric Prieto, "Phenomenology, Place, and the Spatial Turn," in Robert T. Tally, Jr., ed., *The Routledge Handbook of Literature and Space* (Abingdon, UK: Routledge, 2017), 60–9.

36 David J. Bodenhamer, John Corrigan, and Trevor M. Harris, "Spatial Narratives and Deep Maps: A Special Report," *International Journal of Humanities and Arts Computing* 7, no. 2 (2013): 170–227.

37 Another categorization of deep maps may be found in Lotte van der Molen, *Mapping Deep Maps: How Deep Mapping Practices Can Change Our Relation to the City* (MA thesis, University of Utrecht, 2020), 33–54. http://dspace.library.uu.nl/handle/1874/399294. Her types include historical, poetic, playful, collaborative, and political.

38 http://digihist.de.

39 The most comprehensive listing of historical GIS projects can be found at The Historical GIS Research Network www.hgis.org.uk/resources.htm (accessed September 27, 2020).

40 https://culturalatlas.sbs.com.au/ (accessed September 27, 2020); www.literaturatlas.eu.

41 https://artsandculture.google.com/.

42 https://riches.cah.ucf.edu/.

43 www.medievalchester.ac.uk.

44 Vision of Britain. www.visionofbritain.org.

45 Todd Presner, David Shepard, and Yoh Kawano, *HyperCities: Thick Mapping in the Humanities* (Berkeley: University of California Press, 2014).

46 www.enchantingthedesert.com.

47 http://ckcc.huygens.knaw.nl.

48 http://republicofletters.stanford.edu.

49 http://pelagios.org/.

50 Keith Thomas, "A Highly Paradoxical Historian," *New York Review of Books* (April 12, 2007): 56.

51 Stuart Eve, "Augmenting Phenomenology: Using Augmented Reality to Aid Archaeological Phenomenology in the Landscape," *Journal of Archaeological Method and Theory* 19 (December 2012): 582–600.

52 www.smithsonianmag.com/smart-news/how-virtual-reality-helping-prosecute-nazi-war-criminals-180960743/.

53 Kyungyoon Kim, Bret Jackson, Ioannis Karamouzas, Moses Adeagbo, Stephen J. Guy, Richard Graff, and Daniel F. Keefe, "Bema: A Multimodal Interface for Expert Experiential Analysis of Political Assemblies at the Pnyx in Ancient Greece," 2015 IEEE Symposium on 3D User Interfaces (3DUI). https://doi.org/10.1109/3DUI.2015.7131720.

54 Roger Louis Martínez-Dávila, Paddington Hodza, Mubbasir Papadia, Sean T. Perrone, Christoph Hölscher, and Victor R. Schinazi, "Telling Stories: Historical Narratives in Virtual Reality," in Jennifer E. Boyle and Helen J. Burgess, eds., *The Routledge Research Companion to Digital Medieval Literature* (New York: Routledge, 2018), 107–30.

55 Trevor M. Harris, L.J. Rouse, and Sue Bergeron. "Humanities GIS: Adding Place, Spatial Storytelling and Immersive Visualization in the Humanities," Michael Dear, et al., eds., *Geohumanities: Art, History, Text at the Edge of Place* (Abingdon, UK: Routledge, 2011), 226–40, quote at 226.

56 Nicolas Suzor and Darryl Woodford, "Evaluating Consent and Legitimacy amongst Shifting Community Norms: An EVE Online Case Study," *Journal of Virtual Worlds Research* 6, no. 3 (September 2013): 1–11; Bobby Glushko, "Tales of the (Virtual) City: Governing Property Disputes in Virtual Worlds," *Berkeley Tech Law Journal* 22, no. 1 (January 2007): 507–32.

57 Barney Warf, "Deep Mapping and Neogeography," in Bodenhamer, et al., *Deep Maps and Spatial Narratives*, 134–49.

58 John Corrigan, "Genealogies of Emplacement," in Bodenhamer, et al., *Deep Maps and Spatial Narratives*, 63.
59 William James, *A Pluralistic Universe* (New York: Longmans, Green & Co., 1909), 212.
60 William H. Sewell, *Logics of History: Social Theory and Social Transformation* (Chicago, IL: The University of Chicago Press, 2005), 100.
61 Sewell, *Logics of History*, 111.

2 The art of deep mapping

Denis Wood

Pick up a map. How many stories can you tease out of it? Hundreds if the map's any good, thousands if it's a rich topographic survey sheet. The stories lie latent in the things on the map. They can be assembled in any order required. That's the difference between a map and a *deep* map. A deep map *tells* the story, or stories, it was created to embody.

Let me start with a simple example (Figure 2.1). Some people may not want to call this a map. They'd prefer to call it a plan, but what's the difference? It's the floor of the big gallery in the Brattleboro Museum and Art Center in Brattleboro, Vermont. It used to be the lobby of the train station the museum's in. This is where, in 1990, I mounted *The Portrait*, an exhibition that culminated the museum's two-year *Our Town* program, intended to reintegrate the museum with the people who paid for it. An introductory text on the map's other side, wrapped around an empty frame, said:

> Whether painted or taken with a camera, written in music or in words, most portraits consist of thousands of small marks or gestures—bits of pigment or sound, flecks of emulsion or ink—working together to convey, at some other scale, a coherent image. This portrait is no exception. Thousands of individual *objects*—photographs, maps, family trees, antiques and contemporary artifacts, video tapes, oral histories, yearbooks, drawings, plans, models, air photos, panoramas—have been welded together to create, at another scale, a coherent image. The difference is that this image is a frame, a frame for the portrait visitors paint or draw or snap as they not only "put together" what's here, but to which they contribute—literally—through interactive displays. This frame encourages a dialectic view of Brattleboro, first by opposing a view from *within* to a view from *without*, then by setting in tension a view from the *past* with one of the *future*. "Brattleboro," the town the visitor paints, is suspended within this frame. Neither determined by an artificial past nor prescribed by a single future, neither absorbed in the insider's view nor lost in that of the outsider, the real Brattleboro remains . . . somewhere else.

This frame very much wanted to be experienced in a particular order, first the introductory materials, then the view from within, next the view from the past, then the

DOI: 10.4324/9780367743840-2

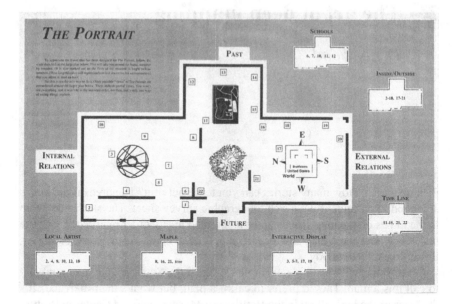

Figure 2.1 The map inside the Brattleboro Museum and Art Center's exhibition guide, *Town Treasures 8: The Portrait* (1990). I wrote the introductory text on its other side and the block of explanatory text here.

view from without, and finally the future, which visitors would encounter—and begin creating—as they left the museum.

The view from within—internal relations—consisted of a number of elements. One of these was the 1,210 photos professional photographer Bob George had taken in Brattleboro over the years. These completely covered the "internal relations" wall (we provided a telescope for viewing images above eye level). This not only called into question the unitary feeling of "portrait" but also began to sketch the genuine richness of Brattleboro life. Another was a very large ring mounted on a Lazy-Susan-kind of base. Visitors were to fasten name tags to this, in alphabetical order, and then to run strands of colored yarn to the name tags of others to whom they were in any way related (by blood, friendship, obligation, or neighborliness), attaching tags for these people when they weren't already there. This not only tied visitors *into* the exhibition but also embodied the idea that cities ultimately consist in their connections. Beyond this lay other interactive "sculptures" that explored themes of family, church, and school (here kids wrote their names on leaves they attached to their teachers' branches of a tree, to which teachers were attached by ribbons to their teachers, and so on). The past was dominated by a wall covered with 600-some historical postcards that had been collected by William Flemming, overlooking banks of oral histories that could be listened to, videos that could be watched, historic sugar-mapling artifacts, and the like. "External relations" was dominated by a wall-size aerial photo of Brattleboro, and a cascade of chains hanging from the ceiling with Brattleboro in their center. Again visitors attached themselves to the exhibit, tying a tag to the Brattleboro chains, and then yarn out to tags they'd attach to chains at the distances of the rest of the state, the nation, the rest of the world—wherever connections subsisted. Now one could see that although Brattleboro consisted of a web of internal connections, it was simultaneously suspended in a web of external connections; and visitors were encouraged to bring in and fasten to the relevant chains "envelopes or tea bags, greeting cards or bills of lading, 'Made in China' tags or movie tickets, long distance phone bills or shopping bags." By the time the exhibition closed this had become an impenetrable mass embodying Brattleboro's place in the world, as the internal relations ring had become a thick mat of local connections.

Our map (Figure 2.1) imposed an order, from the intimate local, through the past, to the world at large and finally the future, which gave the exhibition of heterogeneous stuff a palpable structure, *that let it tell a story*, a story that pushed attendance well beyond historic highs. People loved the show. Maps are useful because they're open, but people love stories, and that's what deep maps give them. This may be a startlingly shallow example of a deep map (though the exhibition to which it was an introduction was *very* much deeper indeed), but it lays out one dimension of the deep map—its directed quality—in a straightforward way. This is an essential quality.

Deep maps owe their name, and a lot of their inspiration, to William Least Heat-Moon's *PrairyErth: A Deep Map*, which he published in 1991, but the endeavor, a deep, close reading of place, is much older, with antecedents reaching back to Gilbert White's *Natural History and Antiquities of Selbourne*, to Henry David Thoreau's *Walden*, to Wallace Stegner's *Wolf Willow*. Fellow travelers include the English topographers of the eighteenth and nineteenth centuries, French psychogeographers of the 1950s and 1960s, and contemporary English psychogeographers like Iain Sinclair (e.g., *London Orbital*). But I too, a geographer with deep roots in art, literature, and maps, had fallen into a similar practice, with broadly equivalent commitments. In fact, I owed my involvement with Brattleboro to a deep map I'd been working on in Raleigh, North Carolina, where I'd found myself on the landscape architecture faculty in the School of Design at North Carolina State University.

Wherefrom

That I'd ended up there was due to an earlier series of engagements that begin my first semester in graduate school, at Clark University's School of Geography. This was in 1967. As an undergraduate I'd triple-majored, entering college to study Medieval history, moving on to geography (originally to fulfill a lab science requirement), and finally to English, in which I graduated. I applied to graduate schools in all three subjects and ended up in geography because Clark offered me the most money. There, in my first semester, Martyn Bowden introduced us to the work of J. K. Wright, pointing with particular emphasis to Wright's essay, "*Terrae Incognitae*: The Place of Imagination in Geography."[1] Here Wright introduced the idea of *geosophy*, "the study of geographical knowledge from any and all points of view." Wright meant "any and all," insisting that he was talking about the geographical knowledge of fishermen and businessmen, Hottentots and Bedouins, kings and kids. Reading this I wondered, why not the geographical knowledge of a sixth grader, a sixth grader who'd just moved from a housing project on the West Side of Cleveland to a strip of apartments in Cleveland Heights? Me, in other words. No doubt I was also homesick—being at Clark was the first time I'd lived away from home—but it wasn't the first time I'd thought about this. When we'd moved to Cleveland Heights we kids rode in the closed-in back of the moving van—for the thrill—so that when they opened the doors and let us out it was almost like, I don't know, being born again. There was light snow on the ground. Everything else was new.

And then, little by little, it wasn't.
How did *that* happen?

That is, how did cognitive maps get built? In my doctoral work we would collect maps from kids every other day, so we were able to track this, *were able to watch their memories being built*, and I'd hoped it would be a "speeded up" version of what I'd wanted to explore here about the growth of my knowledge of Cleveland Heights. Was it? I don't know, but in the event I make a bunch of maps. The first three were attempts to map my earliest impressions: the apartment building, its driveway, the garage and backyard, the walk to the sidewalk and street, the facades of the apartment buildings we could see, a playground. There are notes like, "I knew the street continued but no more," and "discriminate between areas known and noticed, *known*: my apartment, sidewalk to playground, playground; *noticed*: the rest." Clearly this was an exploration of my memories, and fairly deep memories at that, but memories were all I had. Three more maps explored my increasing awareness of the walk to school, this variation, that. Next came a couple that tried to get at our explorations, mine, my brothers' (we were inveterate explorers). And then I made a map to update the map of my first impressions (Figure 2.2). I called this "Second State Overlook Impressions Organization" because Overlook was the name of the road our apartment was on. I guess it was the name of the neighborhood too. In a highly directed way this map tried to discriminate places where I felt at ease—which are blue in the original—from those where I didn't, either because we could be hassled by janitors or because I was intimidated. I was mapping a bigger area too. This was no longer the space in front of my apartment, but some four or five blocks. In the end I made, depending on how you count, 20 maps, all sketches, one of them as late as 1976, none of them ever polished. Four of these went even further back into my memories, to explore my walks, on Cleveland's West Side, from our apartment there to my elementary school.

But I never quite knew what to do with them. They just hung around in my files until 2012 when Becca Hall and Cara Bertron asked me to contribute something to the first issue of their *Pocket Guide: to the known world*. I wrote "Thinking about my paper routes," which had been the focus of my 1976 map, and I illustrated it

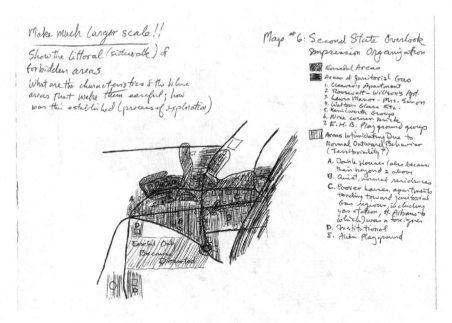

Make much larger scale !!

Show the littoral (sidewalk) of
forbidden areas
What are the characteristics of the blue
areas that make them easeful; how
was this established (process of exploration)

Map #6: Second State Overlook
Impression Organization

▨ Easeful Areas
■ Areas of Janitorial Gas
1. Cleaver's Apartment
2. Roosevelt – Wellson's Apt.
3. Lawn Manor – Mrs. Simon
4. Walton – Glass Etc.
5. Kenilworth Groups
6. Nice corner buckle
7. E. H. B. Playground group

▤ Areas Intimidating Due to
Normal Outward Behavior
(Territoriality?)

A. Double Houses (also because
their beyond 2 above
B. Quiet, normal residences
C. Poorer houses, apartments
tending toward Janitorial
Gas regions, including
gas station, H. Albans to
which I was a foreigner
D. Institutional
E. Alien Playground

Figure 2.2 Second State Overlook Impression Organization was the sixth in the series of
maps of Cleveland Heights I drew in 1967 inspired by J. K. Wright's geosophy.
The map organizes the neighborhood along a comfort dimension as it probes
for the sources of ease. In the original the easeful areas are in blue.

with the map (Figure 2.3).[2] Later that year the publishers at Visual Editions asked
me to contribute something to their crazy box book, *Where You Are: A Book of
Maps That Will Leave You Completely Lost,* and I decided to pursue the paper
routes. This time I asked my brother, Chris, with whom I'd shared my first route,
and Mark Salling and John Bellamy, both of whom had had routes in the neighborhood at the time I had, to add their two cents. After all, they'd all attacked my
memory in the original piece in *Pocket Guide.* "The Paper Route Empire," by me,
John Bellamy and Mark Salling came out the next year.[3]

The book box contained 16 individual booklets, by Alain de Botton, Geoff
Dyer, Olafur Eliasson, and others. My booklet included four of my maps from
1967 and the one from 1976, along with a big foldout map by Salling (Figure
2.4) and two from Bellamy (as well as a paragraph from my brother). And okay,
none of this *solved* the problems of our memories, and the piece *certainly* didn't
end up answering the question how I came to know my Cleveland Heights (that's
still open). But the paper routes were completely tangled up in our lives, in the
emotional lives of who were, after all, teenage boys, teenage boys growing up in
the 1960s, if that matters (I think it does), and *however* I came to know Cleveland
Heights, this had to be a big part of it. (I'd had the routes for 10 years.) "The Paper
Route Empire" is pretty damn deep, and an interesting model for the exploitation
of memory, the exploitation of memor*ies*, in deep mapping.

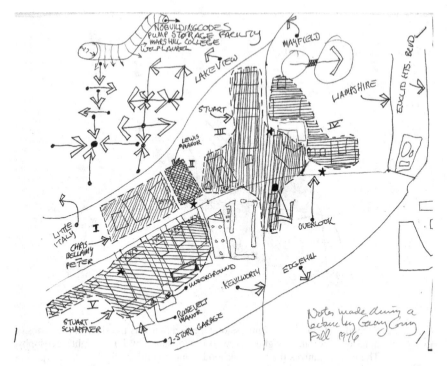

Figure 2.3 This is the map of Overlook I made in 1976 showing the location of my paper routes.

Source: From Becca Hall and Cara Bertron, eds., The Known World, Pocket Guide, 2012, unpaginated.

I went on to explore how cognitive maps get built in my thesis and dissertation, the former of which quite incidentally approximated a deep map. *Fleeting Glimpses; or Adolescent and Other Images of the Entity Called San Cristobal las Casas, Chiapas, Mexico* is an intense analysis of the "mental maps" drawn by 300 high-school students in San Cristobal, a city I'd visited constantly over the previous 10 years.[4] I'd fallen in love with San Cristobal my first night in the Hotel Español—the blankets on the bed, the fire in the room—and it never abated. I loved Zinacantan—I had sort of a second family there—and Mitonic and other villages in the surrounding highlands, but most of all I loved las Casas itself. I went back again and again. I knew the place intimately. I'd been in its city hall but I'd also spent a night in its jail. I'd walked every street. I'd been in many of the homes. As the first sentence of my thesis put it, "Love and long association with a city make it both extremely difficult and all too easy to write about," and I had the responses to 300 questionnaires, long questionnaires that sought lots of answers, many of which had to take the form of maps. I submitted the thesis in 1971 and, promptly published by the Clark University Cartographic Laboratory, it gained a certain notoriety. There were maps galore here (Figure 2.5), and if this wasn't a

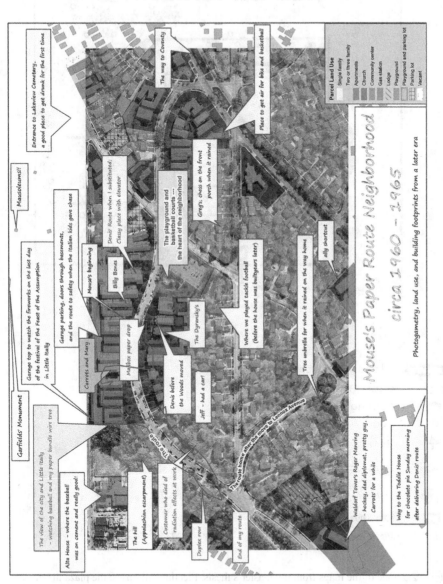

Figure 2.4 Mark Salling's map of his paper route and environs.

Source: From "The Paper Route Empire" in Visual Editions' Where You Are: A Book of Maps That Will Leave You Completely Lost (London 2013).

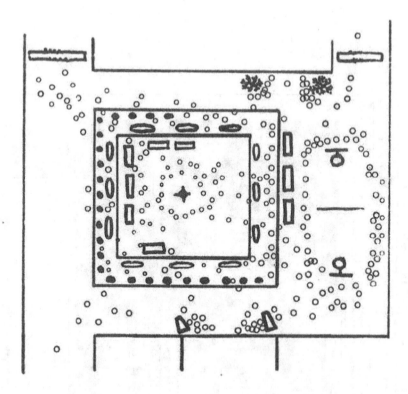

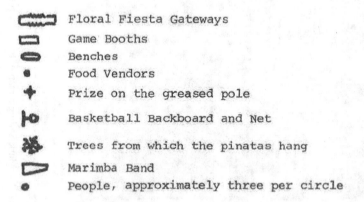

Floral Fiesta Gateways
Game Booths
Benches
Food Vendors
Prize on the greased pole

Basketball Backboard and Net

Trees from which the pinatas hang

Marimba Band
People, approximately three per circle

Figure 2.5 One of the maps from my master's thesis, Fleeting Glimpses: the plaza in San Cristobal's Barrio Mexicanos at 4:00 p.m., August 14, 1969 during the fiesta in honor of the Ascension of Mary.

deep map, it came close, one that suggested how directedness and memory could be supplemented by questionnaires and observation.

My dissertation, *I Don't Want To But I Will*, was legitimately notorious and, while in no way a deep map, it more or less got me the job at North Carolina State . . . where I became a deep mapper.[5] It was 1974—I'd spent the previous couple of years teaching high school—and this was my first experience teaching college. The fall semester was adequately unnerving, but the spring semester promised to be calamitous: I had been assigned an undergraduate landscape architecture studio to teach. Studios are the primary from of instruction in design schools, settings in which students work at drafting tables on projects set by the studio master, the teacher, me. Classically I was to set the problem and then wander among the tables, instructing and commenting on the students' work. Periodically the students would put their work up and we'd all comment on it. (We called these crits.) Then back to work. I was desperate: *I knew nothing about landscape architecture.* But I did know something about making maps and I thought: why not use mapping as a way of selectively focusing the students' attention on those aspects of the landscape that, in the instrumentality of their training as future professionals, they were apt to overlook: the way the land smelled, the way it felt in their legs when they walked it, the sound of the wind in the oaks after all the other leaves had fallen (Figure 2.6)? The university was surrounded by neighborhoods, including mine, and we mapped one of them. It turned out to be a lot of fun and gave me an opportunity to explore weird modalities of mapmaking (out of food, with different textures, etc.) while simultaneously exploring my new environment, practically with a microscope.

I kept giving this studio until 1982 when I suggested to Robin Moore, who lived across the street from me and with whom I was to co-teach the studio that semester, that we aim for an atlas this time, that is, for a coherent *bunch* of maps that at the end of the semester we could photocopy and distribute to our Boylan Heights neighbors.[6] This meant that the students would have to discuss what they wanted to map, and figure out who'd be doing what. After wandering around the neighborhood, we landed on some surprising topics: the sewer system, the gas lines, the water mains, the electric grid, the phone system, the trees, the mail route, the stars overhead, the fences, the graffiti, the night lights, yards, property ownership, oh, yeah, and the streets! The streets became a problem, a serious problem. Of course we made a map of them, measuring them to get them right, but otherwise we couldn't get rid of them. No matter what the subject was, the streets always came along for the ride: *what would the neighborhood be without the streets?* What, I responded, would be it be without houses, water mains, trees, night lights?

What, in other words, if the mapmaking were an *expressive art*, a way of *coming to terms* with place, with the experience of place, with a love of place? I'd never been able to do the wandering-among-the-tables routine, so pretty much we had crits every day. This day we were dismantling a map of streetlights, together, as a group, and we just kept paring away the inessential, the map crap (the neat line, the scale, the north arrow), the neighborhood boundaries, the topography,

Figure 2.6 The color of the leaves in Boylan Heights in the fall, 1982, with the typefaces used in the version that appears in Everything Sings (Siglio Press, Los Angeles, 2010).

finally the streets: first the scaled streets, then a schematic grid of the streets, finally even a hint of a grid of the streets. None of this was easy. Daylight finally went too—that default daylight that most maps take for granted—so that we were fooling around with circles of white on a black background. This made it clear that the map wasn't about the lamp *posts*, but about the lamp *light*, and light was something we weren't sure how to deal with. Certainly, the uniform white circles we'd been drawing caught nothing of the way the light frayed away at the edges; and one night, armed with a camera, we scaled a fence and climbed a radio tower on the edge of Boylan Heights hoping to catch the night lights on film. What a disappointment! The view from above was *nothing* like walking in and out of the pools of dappled light on the streets below. But I had a pochoir brush at home and when Carter Crawford—who'd put himself in charge of atlas graphics—used it to draw the circles it was magical (Figure 2.7). Nothing but blotches of white: that *was* the way it felt to be walking the streets at night!

The usual "efficient" map would have located everything on the street onto a single sheet, that is, different marks for lamp posts, fire hydrants, street signs, trees. Our *inefficient,* but highly directed map, concentrated on a single subject, and rather than the lamp *posts* it brought the *pools of light* into view. No legend, no north arrow, no neat line, none of the usual apparatus. Instead, a sense of poetry, on a map attentive to the experience of place.

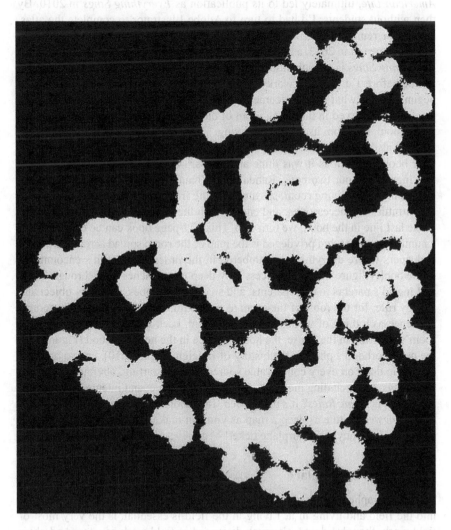

Figure 2.7 The map of the streetlights Carter Crawford drew with my pochoir brush (Everything Sings, Siglio Press, Los Angeles, 2010).

An ever-smaller number of students and I worked on this atlas for the next couple of years. From the beginning we'd explored a range of data sources and this expanded. We made maps from remembered data, observed data, data from questionnaires, archival data from the city, county, state, and various service providers (the gas company, the electric company). We explored different ways to make maps: we drew them with pens and pencils, we painted them, we rubbed things for them, we collaged them. With very primitive software we made the tree maps on a mainframe computer, outputting the maps to a flatbed printer.

Then, as so often happens with my projects, it ended up in a box in the back of a closet. A chance mention to Ira Glass, who was interviewing me for *This American Life*, ultimately led to its publication as *Everything Sings* in 2010. By then without students, I'd had to turn to Adobe Illustrator to complete the atlas, making or remaking half the maps on the computer. An enlarged and enriched edition came out in 2013, and today there's an electronic version you can stream.[7] *Everything Sings* is an authentically deep map!

But before I'd finished working on the atlas with my students, I'd already begun work on what would become *Home Rules*. This was an analysis of the rules my family observed in the living room of our home, well, of the rules, the things in the room, the room itself, and the living. Though I coauthored it with Robert J. Beck, it also has the names of my wife and sons on the title page because it was their book as much as it was mine and Bob's.[8] What Bob did was to interview me, Ingrid, and our two boys, Randall and Chandler, about the rules attached to everything in the living room, starting with the front door, wandering through all the furniture and accessories, and ending with the switches for the lights (which, in the last line in the book, we turn off). This 329-page book can be entered along a number of routes, but privileged is the map of the room spread across the book's endpapers where everything is numbered in the order in which it's encountered in the book (Figure 2.8). This excessively deep map reaches back through Denis and Ingrid's parents to their parents, and so on, as it probes, object by object and rule by rule, for the roots of the room as an institution; as well as for the room's existence as a thing of wood, plaster, and paint, back through the history of the room's situation in the house, the house's place in the neighborhood (Figure 2.9), the neighborhood's place in the history of Raleigh (Figure 2.10), and so on. This deep map drew on every conceivable source of data, starting, obviously, with the interviews, but including memory, every useful archive, and intense observation. How deep is *Home Rules*? It's damn deep. Taken together *Home Rules* and *Everything Sings* constitute as deep a map as you can make. There's an awful lot you can do without buying an airplane ticket!

Settling down with place

I'm a geographer so for me what these deep maps are all about is getting out into the field and living in it. Living in the field is essential, is the very most of field work, though I doubt it's much discussed in field work courses. And obviously when I say "living in the field" I'm talking about, yes, getting out and

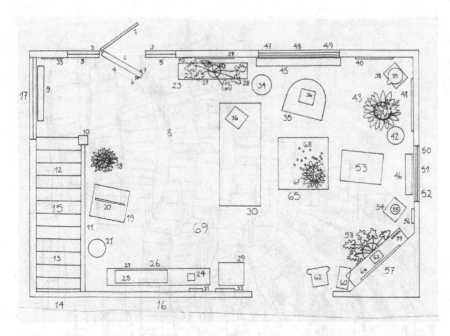

Figure 2.8 The map of my living room at 435 Cutler Street, which runs across the endpapers of Home Rules (Denis Wood and Robert Beck, Home Rules, Johns Hopkins University Press, 1994). The objects in/of the room were ordered in the book as numbered here.

tramping around (a lot) but also about asking questions, formally as well as informally, about raking through archives of every imaginable variety, and heaven knows about reading very, very widely. It's not evident in most of my work how widely I cast my net, but in *Everything Sings*, for example, the simple maps "Absentee Landlords" (Figure 2.11) and "Local Rents"—which explore property ownership—could only have been made after spending hours in the offices of the Wake County Register of Deeds where we ran ownership back to William Montfort Boylan whose death in 1899 led from his estate to the creation of Boylan Heights. From there we created a tree of descent to the ownership pattern present in 1982. It was a mammoth amount of work and generated a drawerful of cross-referenced file cards. The map of "Assessed Value" involved a similar tedious labor, though I made the map itself (Figure 2.12) by smoothing the tax values of 333 residential lots across a 281-cell grid and interpolating contours across the grid. Then I glued strings to the contour lines and made a rubbing of the strings. Today, of course, all these data are available online and often have already been mapped, but in 1982 that was very much in the future.

On the other hand, the map I made for an article about the atlas for *Places Joural*, "Porch Ceiling Colors," could only have been made by peering up at the porches' ceilings from the sidewalk.[9] This was also a riposte to an observation

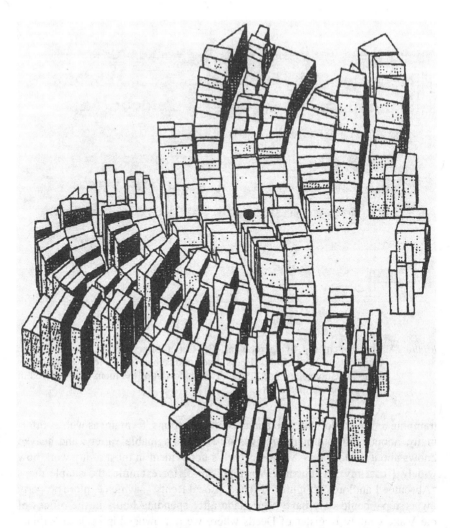

Figure 2.9 The place of our house (with the dot on it) in Boylan Heights, where each
oblong indicates, in its plan, the shape of the lot it's on; while its height
indicates its conformance to the Boylan Heights norm, taller being closer to
that long, narrow lot with a bungalow or shotgun on it.

Source: This is from Home Rules, but another version also appears in Everything Sings.

by an aficionado of remote sensing data who claimed he could have made the
atlas without leaving his desk. Part of our field work had consisted of walking
the neighborhood house by house and noting everything we thought might be
useful for making maps. Among these were the colors of the front porch ceilings. We hadn't originally made this because the map had to be in color and the
atlas was black and white; but once de rigueur in the South, and reportedly still

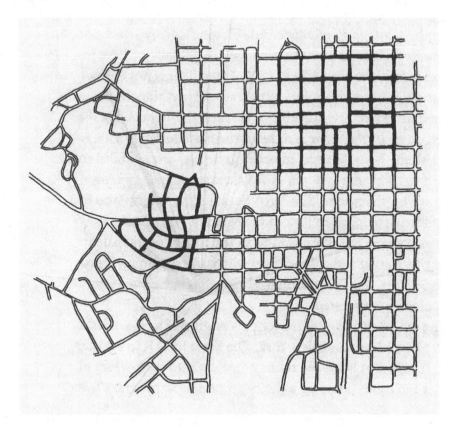

Figure 2.10 The place of Boylan Heights, curvy streets, darkened, left center, as laid out at
the beginning of the twentieth century, in the context of downtown Raleigh's
orthogonal street plan, the late eighteenth-century core, darkened, upper right.

Source: This is from Home Rules, but similarly it also appears in Everything Sings.

predominant, blue porch ceilings were but a small minority in Boylan Heights (at
least in 1982). Most ceilings were white, but green and even yellow were more
common than blue. When your ceiling wasn't white, people in Boylan Heights
noticed.

For us there were clearly maps that demanded immersion in archives of every
variety, but just as certainly others couldn't have been made without wearing out
shoe leather. If your field is a couple of hundred years old, patently you're going
to be spending a lot of time in libraries. But contemporary or ancient both are
fields, arrays of data in which you've got to *immerse yourself*. I suppose fieldwork can be parsed into phases, though I hesitate to make it sound more organized than I want it to. But it *has* to start with *getting into the data*. That's the
essential, I-can't-emphasize-this-strongly-enough part of it, getting into the data.

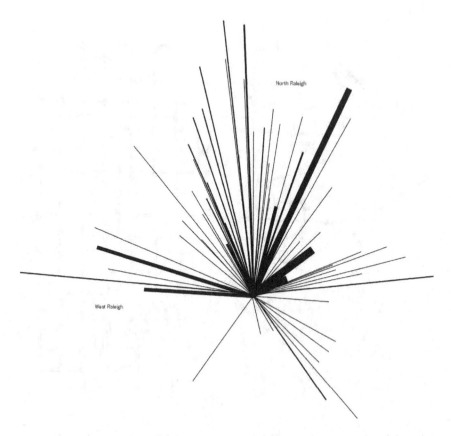

Figure 2.11 This is a map of the "Absentee Landlords" living in Raleigh. Those elsewhere in the state and nation showed up on smaller maps on the facing page (Everything Sings, Siglio Press, Los Angeles, 2010).

"Embark on a garden with a vision but never with a plan," Ian Hamilton Finlay wrote, and it's as a good maxim for our kind of work as I can imagine.[10] Embark with a vision . . . You don't even want the vision to be too heavy-handed but god knows *you don't want a plan*. What you're doing is keeping yourself loose, open, absorbing, settling into the material, with the vision lightly directing your gaze. Lightly. What you want to be able to do is *see*, but also to be able to shift your gaze if that's what the data are . . . suggesting? promoting? encouraging you to do? You're relaxed, but you're alert. I could say this 60 different ways, but the bottom line is: no plan!

The thing is, deep mapping is an *experiential* science, not an experimental one. It's about *finding facts*, not creating them. Plans restrict experience. If you're looking for the Higgs boson, oaky, you need a plan, just to think about getting started looking for the Higgs boson you need a plan. But even if you imagine that

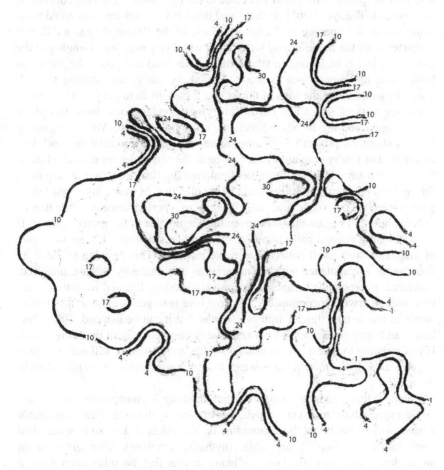

Figure 2.12 This is the map I made of "Assessed Value" in 1982. The numbers were in thousands. This was a long time ago. They're completely out of sight today (Everything Sings, Siglio Press, Los Angeles, 2010).

the search for subatomic particles only began in the nineteenth century, humans had been mucking around in that world for a *long* time, keeping their eyes peeled, wandering down this dead end and that, before beginning to make plans to look for stuff. And if it's deep mapping you're wanting to do, you're not going to be looking for subatomic particles. No, the rules for experimental science have little place here. Deep mapping is about *the experience of place.*

So you're out in the field—and this could be sitting on the terrace of a cafe watching the world go by or immersed in a sea of marriage certificates in some archive—and where you want to head is beginning to come into focus. This is when you begin to sense the *direction* you're heading, *the direction you want to take your deep map.* The vision you came to the field with is growing concrete—you could do this, you could do that—and this kind of thinking (you could think about this as the opening of the second phase of the fieldwork) has a different character from the thinking you were doing when you were just lounging in the data. A project is beginning to take form in your head and you're beginning to think about its shape, about its scope. As you do, you're also starting to think about the work, about the sorts of things you'll have to do to get you what you're beginning to think you might need. I'm trying to suggest how loose this phase is, how open you are to other directions, other possibilities. You're beginning to think about the project's final form, mind-casting the results of the work into that form, but you're continually feeding back the consequences of each of these thoughts into any and/or all the earlier thinking. The final form? For a deep map, for *my* kind of deep map, this should be a collection of maps, but again this is going to depend on you. It *could* take the form of a performance on kettle drums.

Now: how to imagine the work given this vague idea of the project's form? If we're thinking about a collection of maps—and let's do that—it's got to consist of mappable stuff, stuff collected to make maps of. This is going to entail an ever-deeper acquaintance with the material as it is endlessly walked, analyzed, measured, photographed, rubbed, reviewed, collected, listened to, tasted, sampled, and otherwise encompassed. The goal here is to get as close to the stuff as possible, and to take this all in in such a way that it can be mapped. This is less hard than it may sound, but it's not necessarily easy and it *is* all time consuming. If you're going to give a performance on kettle drums, you're still going to need to get as close to the material as you can, though it won't be with the goal of making maps.

Crucially this means not imagining that anything is unmappable, not percussively expressible. But I know maps. Let me stay with maps. There's probably a better way to put it, but the bottom line is: if it exists, it does so in space. And pretty much, *therefore it's mappable.* Anything. Sometimes this might mean an unimaginable amount of fieldwork, of being on your feet, but other times it might require you to sift through reams of paper. *Imagination is key.* Invisible radio waves? Well, they're waves, they have form, they're emanating from propulsive forces somewhere. Locate the radio sources, the towers (Figure 2.13), and calculate the relevant wave fronts. Draw them. There's your map (Figure 2.14). But at other times it really is just plodding through the data as often and as *slowly* as necessary. And then you just have to commit everything to as many maps, or musical scores, or whatever as it takes. Collaborative prose helps too. And anything else.

And if, as you make these things, you realize you're missing something, then it's back to the data, back to the field. The scope's constantly changing, its form is morphing, the project is changing as you engage in the work of discovery and of explaining to people what you're finding. It should change your life. Because it should feed back into the place, even if not extant, it should change the life of the place moving forward. Deep mapping is a deep, highly directed, multifaceted engagement with place. If it's not, it's not a deep map.

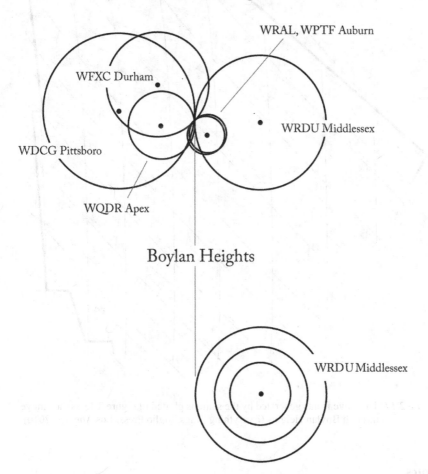

STATIONS Transmitting from

WRAL, WPTF Auburn

WFXC Durham

WDCG Pittsboro

WRDU Middlessex

WQDR Apex

Boylan Heights

WRDU Middlessex

Radio waves spreading from WRDU in Middlesex

Figure 2.13 Boylan Heights is at the intersection of these wave fronts being generated by
the towers indicated. The lower image indicates the shape of the WRDU wave
front by the time it reaches the neighborhood (Everything Sings, Siglio Press,
Los Angeles, 2010).

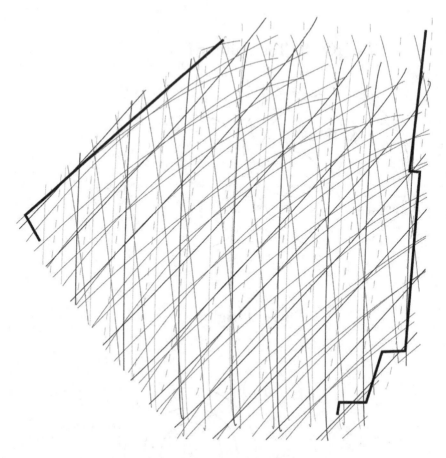

Figure 2.14 The wave fronts generated by the stations plotted in Figure 2.13 as they move through Boylan Heights (Everything Sings, Siglio Press, Los Angeles, 2010).

Notes

1 Martyn Bowden, on the faculty at Clark, was a huge Wright advocate. He encouraged me to buy, and read, J.K. Wright, *Human Nature in Geography: Fourteen Papers, 1925–1965* (Cambridge, MA: Harvard University Press, 1966), where the one about geosophy is "*Terrae Incognitae*: The place of imagination in geography."
2 "Thinking about my paper routes," appeared in Becca Hall and Cara Bertron, eds., *The Known World, Pocket Guide* (2012), unpaginated, the first issue in their "foldy" project, a single sheet, folded, and sent through the mails from Seattle.
3 Denis Wood, John Bellamy, and Mark Salling, "The Paper Route Empire," in *Where You Are: A Book of Maps That Will Leave You Completely Lost* (London: Visual Editions, 2013), one of sixteen booklets (22 pages, fold-out map). It's online at http://where-you-are.com/denis-wood. A longer, online version appeared as "The Paper Route Empire," *Belt Magazine* (November 11, 2013). http://beltmag.com/paper-route-empire/.

4 Denis Wood, *Fleeting Glimpses, or Adolescent and Other Images of the Entity Called San Cristobal las Casas, Chiapas, Mexico* (Worcester, MA: Clark University Cartographic Laboratory, Clark University, 1971) is out of print but can be found online at www.deniswood.net.
5 Denis Wood, *I Don't Want to, But I Will* (Worcester, MA: Clark University Cartographic Laboratory, 1973). This too is online at www.deniswood.net, though you may also want to see the discussion at www.metafilter.com/117054/I-Didnt-Want-To-But-I-Did.
6 I've written about this in Denis Wood, "Mapping Deeply," *Humanities* 4, no. 3 (2015): 304–18, which is available in print in *Deep Mapping*, Les Roberts, ed. (Basel: MDPI, 2016), 15–29, and online at www.mdpi.com/2076-0787/4/3/304/htm.
7 You can hear the This American Life interview at www.thisamericanlife.org/radio-archives/episode/110/mapping. The atlas appeared as Denis Wood, *Everything Sings: Maps for a Narrative Atlas* (Los Angeles: Siglio, 2010); and in a revised and expanded edition, in 2013. The electronic version of the book is here: http://sigliopress.com/book/everything-sings/.
8 This really emerged from the ongoingness of both our lives, as Bob spent several weeks every summer in the vicinity while his wife worked at the American Dance Festival in Durham. Mulling what do with our time lead to Denis Wood and Robert J. Beck, *Home Rules* (Baltimore: Johns Hopkins University Press, 1994); Denis Wood and Robert Beck, "Do's and Don't's: Family Rules, Rooms, and Their Relationships," *Children's Environments Quarterly* 7, no. 1: 2–14 (1990); and Robert Beck and Denis Wood, "The Dialogic Socialization of Aggression in a Family's Court of Reason and Inquiry," *Discourse Processes* 16, no. 3 (1993): 341–62.
9 Denis Wood, "Everything Sings: Maps for a Narrative Atlas," *Places*, October 2011. https://placesjournal.org/article/everything-sings-maps-for-a-narrative-atlas/.
10 I found this in Jessie Sheeler's, *Little Sparta: A Guide to the Garden of Ian Hamilton Finlay* (Edinburgh: Birlinn Ltd., 2015), 2; but it's from Finlay's *Detached Sentences on Gardening*.

3 Designing for mysterious encounter

Three scales of integration in deep mapmaking

Nicholas Bauch

Introduction

In this chapter, I examine the role of *integration* in making deep maps. Deep maps integrate at three scales. At the macro scale, they integrate two major geographic epistemologies: abstract knowledge of a place (in the form of big data) with embodied knowledge about that place; or, distanced facts with the poetry of lived experience. At the meso scale, they integrate entities, both natural and cultural, into hybrid forms indicative of the Anthropocene. Designing a digital platform such that users are privy to *how* meetings among material and mental entities happen in place is to showcase a geographical application of assemblage theory;[1] these "meetings" are what I am terming the meso scale of integration. And at the micro scale, they integrate various media about a specific place (like maps, videos, and sound recordings) with each other in a single platform. Two other sections augment this chapter: one on the relevance of collage to deep mapmaking, and the other on how argumentation might function in a paradigm based in encounter.

The act of integration, at each of the three foundational scales, is an act of creation. Making a deep map is not setting out to represent reality, but to craft a portrait of place that at once engages the senses and the intellect. To do so requires an approach to design based on encounter. If reading a deep map is being open to suggestion, then making a deep map is to give opportunities for encountering, where rhetoric and position are communicated through the mystery of what content and visual forms might come next.

Macro scale

In her investigations into the history of enchantment in modern life, Jane Bennett reminds us of something Max Weber wrote. She paraphrases him thus: "the enchanted world is always in the process of being superseded by a calculable world, [but] rationalization never comes out even with nothing left over."[2] Weber was suggesting that no matter how much reason you apply to something there will always be parts of it that are magical and mysterious. There are always questions left over, always something new to solve. To think that progress moves toward a state of complete mastery or understanding, therefore, is a fallacy. It is even

DOI: 10.4324/9780367743840-3

possible that the reverse could be true, that the rationalization process tends to produce more and more mystery by continually pushing the horizon of understanding a little farther out.

This "left over" mystery that we encounter as a by-product of enumerating, cataloging, calculating, and analyzing the world in the name of understanding it more thoroughly is an element on which deep mapmakers can focus their design efforts. At the macro scale of integration, deep maps are the application of big data to the study of specific places. This is a move toward knowledge of places that far exceeds the senses, though not one that forgets the senses. Deep maps are the wedding of distanced, abstracted information gathered *about* a place with intimate knowledge conjured *in* place. Though deep mapmakers will surely maintain that complete understanding of place is never actually possible, a unique contribution of any deep map is nonetheless to go "beyond the sensorium" of what one might phenomenologically ascertain in situ.[3]

The translation of tabulated data to dynamic cartographic portraits is the province of neogeography and information visualization, around which design paradigms and debates have blossomed.[4] Making a deep map is different than making a map, though, a difference that is found in its insistence on integration at all three scales. Whereas information visualization focuses on the technics of harnessing data sets coupled with graphic solutions, a deep map needs a design principle divergent from cartography. Integration lies in the purview of method, and user interface design is at the heart of answering "how" one makes a deep map.

Mysterious encounter is a guiding design principle that could help accomplish the integration of data with embodied knowledge in the user interface platform. Intentionally organizing the tidbits of original source material about a place—the traditional process of which is detailed by Umberto Eco[5]—such that users make unexpected connections among the assets is to achieve one of the defining characteristics of a deep map. Note this is quite different than the researcher using a digital platform to perform her own queries in a behind-the-scenes tool.[6] Designing for mysterious encounter is not about using the platform as a researcher's sandbox, but is instead using the platform as a medium of communication and creation. It is world-building, more akin to crafting the scene of a novel than juxtaposing disparate informational assets.[7]

This serves two immediate purposes. The first is that users have the opportunity to experience a deep mapmaker's vision without the pretense that it is a neutral or objective asset integration tool. While a successful deep map ought to invoke honest and critical understanding about the place in question, it need not operate under the belief that objective reality is being recreated. This leads to the second immediate purpose, or benefit, which is that the deep map can be brought into the fold of established digital humanities critique regarding content, technics, style, and authorship. There is no reason for an author to assume design neutrality, but rather the conversation is about how the authors wove together information about place, the worth of the authors' creative interpretation, and the success or failure of the "informed speculation" thereof.[8] Designing for mysterious encounter in deep mapmaking shares with fiction, therefore, an element of speculation. The depth of a deep map is to extend past what can be mechanically positioned on a cartographic plane, and to do so requires the researcher to extend past what is traditionally held as safely defensible argumentation. Inasmuch as informed speculation is part of what deep mapmaking is, the creator of a deep map is mobile on the continuum between historical geography and the visual arts.

Meso scale

At this point it is worth moving into a discussion about why *mysterious encounter* is a theme with broader intellectual value in geography, and why I believe it fits so well in the cradle of the deep-mapping enterprise. Bennett's recalling of Weber—that enchantment in modern life is that which pokes through the dominant blanket of rationalization, a blanket that can never cover and explain all phenomena—has again recently been echoed by geo-humanists Hawkins and Straughan. They write that in geography "the aesthetic is embraced for its sensuous explorations of subjects, bodies and spaces through a focus on experiences that are in excess of rational thought."[9] They suggest that geographers concerned with dramatic changes in modern human-natural systems have turned this extra-rational realm. A prevailing way of reflecting devastating environmental changes has been to articulate how the earth's environments are now characterized by ubiquitous hybridization between biological and inorganic material systems with human technology and artifice.[10] One of the prevailing metaphors for carrying out this work has been Deleuze and Guattari's *assemblage*, a concept not unlike network, but with an added connotation of process, as in there are forces doing the assembling process of bringing together the earth's (not so) latent material with human intention.[11]

Describing the ways in which human and nonhuman entities come together to forge hybrid environments—like Richard Misrach's portrayal of Louisiana's cancer alley—characteristic of the Anthropocene has become an important approach for those who study twenty-first-century environmental issues.[12] This environmentalism has distanced itself from notions of pristine nature, and its practitioners instead try to expose environments for what they are now believed to be, using terms like "freakology," "monsters," or "industrial ecosystems."[13] Exposing and reframing the constitution of anthropogenic nature is a task well suited to the breadth of geographical study—combining physical processes with cultural interpretations thereof—and as such has been taken up by a range of scholars in and out of geography. Doing this kind of geography, though, is often most successful when it is place-based.

When entities come together to make these post-natures, something happens to the material reality of their constitution. As philosopher Graham Harman puts it, "shuffling objects into ever newer and stranger combinations, modernity creates monstrous unions between the most far-flung objects under the sun."[14] What kinds of strange objects are combined in (post-) modern environments? The Rocky Mountain Arsenal National Wildlife Refuge is a frequently cited example in environmental history—a former weapons manufacturing plant in Colorado that, because of its toxicity, is devoid of human habitation. It is, ironically, for the same reason, therefore home to vast animal populations in one of the region's most successful wildlife refuges.[15] Here the buried, hidden detritus materials used to make explosives live alongside deer, owl, and bear populations—a strange and indeed monstrous union.

Importantly, and to the main point, in assembling these seemingly unnatural relationships among objects and people there are constantly occurring *encounters* that do the work of creating new types of natures—or, to put it in the language of deep mapmaking—new types of places. What happens when two non-humans encounter one another? These types of encounters are meaningful if you believe that all the entities of the world—for example, people, glaciers, buildings—are composed by their relation to other entities. As geographer J.D. Dewsbury puts it, assemblage is "a process of putting together, of arranging and organizing the compound of analytical encounters and relations."[16]

What assemblage theorists have not yet put their collective finger on, however, is translating this "process of putting together" into visual forms. They have also not yet developed place as an organizing principle for structuring and representing the assembling process.[17] It is one thing to write about the history of the Rocky Mountain Arsenal, and another thing entirely to visualize the specifics of what moments of "encounter" look like in a spatial format. What does the moment of encounter look like—how is a profoundly new type of nature made in place—when a bear sniffs her nose at a ruined missile shell, or a honeybee gathers nectar from a cultivated field and brings it to a keeper's comb?[18] Looking to the methods and techniques of deep mapmaking might help answer these questions.

Which objects and organisms become enrolled with one another to make hybrid entities has an *uncertainty* to it. If anthropogenic environment is your topic of study, then it makes sense that this uncertainty should find its way into the representations of new natures themselves. Deep mapmakers can intentionally design for this uncertainty such that users witness the encounters generative of hybrid environments. It is the meso scale of integration that brings entities in proximity with one another, and the deep map's adherence to place as a way of knowing is a snug fit with this environmentalist-intellectual project.

Collage and the meso scale

In a post-human world composed of "relations" among people, machines, and objects, the theme of encounter runs strongly. Encounter is also one of the defining characteristics of twentieth-century collage, which is about bringing "assets" into conversation with one another. In 1912, when Pablo Picasso and Georges Braque began making collages, they were practicing an ancient art form—some of the earliest forms of human creative expression involve bringing together varieties of found objects. But these moderns looked to collage as a reaction against a paradigm that saw painting as a "porthole" through which reality was viewed from a distance.[19] Instead of a single plane that was a single representation from a single painter, the collage was tactile, quotidian, heterogeneous, and accessible. In short, it bridged the gap between life and art. The materials of everyday life did the work of art, that is, they became the medium through which society was refracted and contemplated.

One comparison we can draw between collage and deep mapmaking, then, is the synthesis of heterogeneous assets. In early-twentieth-century collage these assets frequently included things like newspaper clippings, rope, photographs, tickets, and the like, which were all brought together into a shared frame. Over one hundred years later, in twenty-first-century digital humanities setting, the synthesis of heterogeneous assets also takes place within a frame, that of the computer screen. In each case the practitioner scours his or her environment for objects (e.g., newspaper clippings or data sets) and makes aesthetically or topically motivated decisions about what of those objects to include and how to arrange them in the frame.

The second shared characteristic between these two media is that they both, in different ways, strive to erase the distance between the viewer and the world. In collage, the artwork was not an approximation of life that conjured reality but was itself made from reality. The role of representation had changed. In painting—at least the painting to which Picasso was reacting—marks of paint on the canvas stood in for a part of reality, cueing the viewer to imagine or remember life events, object, or places. But in collage nothing stood in for reality, and so the thing being represented was much more abstract, a grander gesture, perhaps even "deeper." Collage brought art into the same mode of reality as the viewers, connecting them to a material experience that was more phenomenological than intellectual.

Likewise, to make a deep map one must thrive in the liminal zone between reality and representation, where the line between the two is no longer a line but a mutating shape, where the riches of the body's senses and the mind's knowledge both make one wonder where mediation stops and life begins again. Deep mapping as such contributes to a fundamental humanistic-geographical act; it is an effort to describe places that are not one's intimate own. It is to compile, artistically arrange, to guide, to collage.

Encounter is the thread that ties together the process of assembling digital assets in a deep map with the historic art of collage making. Makers in each of these media carefully select and arrange their objects such that the viewer might gain a sense of wonderment. The link back to Bennett's enchantment is that non-human objects, like discarded weapons, deer, or honeybees in farm fields, cannot "encounter" one another without operating in an extra-rational ontology. There is a remainder, something unaccounted for that, as it turns out, is extraordinarily good at shedding light on the new environmental forms of the Anthropocene. They are slivers of openings into worlds that we do not see because we do not look for them. Nature in the Anthropocene has to be unpacked entirely because the material reality of ecological systems is changing so rapidly.[20] We need to start the research process from an askance view that seems to come from Weber's extra-rationality, and designing for *mysterious encounter* in deep maps is one way to go about this research process, to start believing that what makes no sense together at first can become an ecological truth.

Weber's "remainder" is the feeling, or mode, of enchantment that people tend to attach to places. Seeing the unexpected remainders of rationalization is a matter of seeing unexpected encounters—of objects and events colliding in ways that carry viewers to a deeper understanding of the place in which these encounters unfold. After all, postmodern natures must happen somewhere, and deep maps have the potential to grasp these "somewheres." They can visually articulate how the hybridizing geographical reaches of the postmodern are so radically different from the Arcadian dream of isolated mountain pastures, unaffected by and unconnected to the rest of the world. At their best, contrary to Arcadia, deep maps use place not as a bounded locale, but as a loom that weaves together disparate elements inside a frame.[21]

One conclusion to draw from this line of thinking is that the power of a deep map—and ultimately its usefulness in the study of place—is its use of a single platform to integrate the elements of a place that are often siloed into rational categories of analysis. Drawing on Dewsbury again, the French *agencer*—incorrectly related to the English "assemble"—is more accurately translated "as a way of bringing forth and mapping out a territory at the same time."[22]

For deep mapmaking, "bringing forth" is the process of assembling, or designing a platform that integrates digital assets about a place in a way most conducive to mysterious encounter. If "bringing forth" a territory (or place) is the first part of this provocative equation, then what about "mapping out" that place? This leads to a question about what it means to perform the act of mapping, in particular, when making a deep map. So far, I have focused on the viewer, and designing digital platforms so that he or she can witness the meeting of various constitutive elements definitive of a place. But what from cartography can be applied to the making of a deep map? In Dewsbury's translation of Deleuze's *agencer*, this mapping out of the territory is connected to nineteenth-century colonial projects in which mapping out a territory was to control and restrict the movement of goods, nature, and people, leading to unknown levels of social crises.[23] Drawing precise measurements and exact relative locations on paper—or at least the attempt to do so—tended to erase many geographic realities existing on the ground all over the colonized world. Colonists used maps as *reasons* for political ideology and action (e.g., blank spots), rather than recognize them as *products* of political ideology and action. Deep mapmakers do not need to rely on this part of cartographic history and can build graphic forms for spatial representation that jettison any strict adherence to precise measurement when that precision severs indigenous or alternative political spaces. What these graphic forms look like and how they function with respect to map users is one of the charges that deep mapmaking has opened.[24]

Micro scale

How do you make a deep map? By integrating "heterogeneous elements," just like in a collage.[25] Is it up to the user to wade through the collage of information and make sense of it? On the surface this seems to be what deep maps are good at: "simultaneous representation," but not argument. Text is the medium that most humanist scholars use to convey intellectual distillation. When deep mapmaking is brought into the purview of the humanities, then, a problem surfaces that connects back to the visual excesses of capitalist culture highlighted by the French artist and philosopher Guy Debord. He looked at the overflow of visual imagery in society as an undesirable trait of masking reality—what he called a "spectacle" far removed from reality. For our discussion about argument, therefore, the roots of a challenging paradox were planted.

On the one hand, deep maps are excellent at organizing encounters, inviting users to synthesize the heterogeneous elements of the deep map, presumably in a nearly infinite variety of ways. But on the other hand, I argue it is direct argumentation that can keep deep mapping away from the trap of Debord's spectacle. The problem is that direct argumentation is better suited for the textual medium because, again, it is a distillation, a discovery of a golden nugget of thought that can be explained with words. Academic discourse is founded on this type of communication; the most impactful knowledge has historically most often been that which can be explained with words. The process of making a deep map, then, suggests that we need a new type of direct argumentation. How are visual arguments made? In the history of art, this is one question that seems to follow from Debord's critique.

Paramount here is relevancy; making a deep map is not about piling on details and features until one's technological capacity is maxed out. If this were the case, then a deep map would be synonymous with a technological treadmill, constantly responding to the technology industry's capitalist advances.[26] Integration of media types has become a form of knowledge production in and of itself, and I believe deep mapping will benefit from keeping a critical stance toward this practice. Within digital humanities, examples of media integration include photography, text, and cartography; sound, databases, and videos; and virtual reality, sculpture, and the internet of things. My concern as a digital geo-humanist is that the technics of a project have the capacity to mask its substantive purpose. But what is at the heart of the "technics" of a digital humanities (DH) project—what is doing this potential masking? More often than not, the core of the technical challenges involved in making a DH project stem from the untranslatability from one media type to the next. Translating a database into a cartographic picture—possibly the most straightforward of translations—is still not a given tool in the skillset of DH scholars or even geographers. There is a plethora of GIS tutorial mini courses offered by academic libraries around the world, week-long mapping workshops going on every month, and advanced degree programs offered by hundreds of universities, all for the sake of gaining proficiency in this translation from dataset to map. The point is that if tools as "old" (going back to the 1960s) as GIS-like programs are not automatically a language spoken fluently by spatially minded academics, then how are less common integrations among other media types supposed to happen?[27]

Integrating media types allows us to imagine what else might happen when these two media are joined. While a major coup in mixed media art—deserving all accolades and encouragement—deep mapmakers should not uncritically borrow this endgame. In other words, the criteria used to make value judgements about mixed media art should not be the same as those used to make value judgements about a deep map. This is part of the challenge of deep mapmaking: it is still first and foremost a geographical practice that seeks to make statements about the becoming of place, and—as my own topical urgency—the material constitution of the Anthropocene. Deep mapmaking borrows the approach of the visual arts to achieve this end.

Argument through encounter

In the DH there is an ongoing tension between encounter and argument. Like many tensions in scholarly practice, it is mostly a productive one, where on the one hand DH teams are focused on making pointed statements in their fields, and on the other are keen to take advantage of visual–digital media that allow users to playfully explore. It is by now a longstanding and well-documented tension, with most DH projects exhibiting both of these characteristics in varying dosages.[28] The commitment to make a deep map is parallel to using encounter as a *means* of argumentation in one's final representation. The type of deep mapping advocated for in this chapter is that practitioners take command of encounter and direct it from perceived serendipity to a carefully designed experience. Encounter does not only need to be a random coming together of elements in a digital platform but can also be—as I refer to it here—something that is intentionally assembled by the deep mapmaker. Less a cartographic technician, then, and despite its title, the deep mapmaker is really a designer of semi-fictional, speculatively informed worlds that illuminate one's place of study.

Deep maps implicitly suggest multiple responses rather than explicitly stating an argument for a truth. They are a response to the age of big data, a series of design solutions to a radically expanding base of primary documentation. The way one structures the archive of a deep map yields representations of places that guide users toward potential conclusions about the places themselves. Making the base structure sing—that is, coordinating its emergences to the graphic interface at just the right time based on a user's input—is one of the most important questions one can ask about methods in deep mapping. Deep map content is seldom experienced linearly, and because of this, deep mapmakers operate in a medium where the moments of critique and the creation of knowledge happen much more subtly. Authorial decisions and intentions tend to become masked in the digital experience itself more than they would be in a typical text-printed argument. It is important that a user remain empowered to make decisions about encountering the deep map to keep her aware of the curated—and therefore subjective—nature of the deep map.

Where does one find authorial and/or curatorial position, rhetoric, and argument in a deep map? Because deep maps wed big data with place, they run the risk of appearing like neutral archives. When making a deep map, voice is never left out, but can be harder to pinpoint because readers are generally not accustomed to deciphering the rhetoric of latent archival positions. Deep mapmakers present arguments for something, but to an audience that is not literate in big data wrangling. Even what may appear as a listing of information about one's place of study is always more than a catalog or a digital memory bank; it is imbued with interpretation.

Making a deep map could be interpreted as the big data versions of geographical description, harkening to the ancient practices of travel writers like Strabo of Amasia.[29] However, in his critique of regionalism as a scientific practice (a regionalism in the long lineage of Strabo), Kimble wrote:

> at best, a regional study can be only a personal work of art, not an impersonal work of science—a portrait rather than a blueprint. As such, it can have substantial value, but its value will lie in the realm of illumination and suggestion rather than of definitive analysis and synthesis.[30]

Attaching a graphic user interface to a database takes the deep mapmaker beyond curating an archive, or as Kimble would say, beyond the "blueprint." Argumentation in deep maps is in the way it illuminates and suggests—between the extreme subtlety of an archive (arguments that are hard to detect) and the explicit statement of a text (arguments that are intentionally articulated). It thus—along with other types of visual digital productions—requires a new type of criticality, an interpretive zone that walks the line between provocation and organization of fact. The challenges involved in peer reviewing digital humanities projects are not unrelated and have begun to be documented.[31]

Conclusion: political economy of visual representation

To conclude, I return to Guy Debord and his famous work *The Society of the Spectacle*. Here, Debord introduced the concept of the "spectacle" as a way to critique what he interpreted as a representational layer that blanketed lived reality. As he put it, "life is presented as an immense accumulation of spectacles. Everything that was directly lived has receded into a representation."[32] This historical moment (the 1960s, though no doubt even more intensely since) is problematic for Debord because the "technical apparatus" of representations cannot be neutral; it is part of a mass media machine that only stands to gain from constructing images, replacing tactile encounters with visual experiences either tailored to or hijacked by the non-neutral image makers.

Debord's observations about the role of visual representation in modern economies have been a lasting critical position because of its apparent pervasiveness; examples seem to abound. A billboard seen from the driver's seat of an automobile on a highway reconstitutes what used to be one's tactile relationship with place as distanced and exclusively visual. Visiting national parks is often reduced to a photograph taken near a parking lot with a preformatted scenic overlook.[33] Human socializing is performed ad nauseam on an ever-shifting platform of websites and mobile applications specifically designed to extract data.[34] Television and YouTube stream from cell phones. As Debord would likely say today, these and countless other instances where designed images become the majority of one's direct experience have created a new kind of objective reality defined by commercial interest. To normalize the billboard, the tourist snapshot, sharing online photographs, and the ubiquitous and infinite access to moving images is to live in a capitalist sheen, a reality equally real, yet planted in extractive modes of production and trade. Living in this sheen, or "spectacle," also engenders new kinds of embodiment, less about putting one's body in direct contact with physical–geographical formations and more about the body as a vessel for the seeing eye and participating economic will.

Where does this leave our discussion of deep maps and integration? The sets of productive questions to be gained from Debord are: (1) How, given that deep maps are decisively visual and representational, does a deep mapmaker resist contributing to the political economy of the spectacle? How, through the design process, can one produce a digital sheen positioned counter to the waves of digitization brought on by the profit motives of Silicon Valley and all the tech valleys of the world? Are there any innocent tools and means of delivery of a deep map? (2) How is the deep map user meant to position herself psychologically and corporeally on the continuum of digital material? If a successful deep map evokes a heightened sense of place, could one conceivably be more "in place" by engaging with a deep map than by being there in body? How would one know if they are learning more about a place through its representation in a deep map or by being there in situ? Here it is beneficial to look again to Jane Bennett, who, in her work after *The Enchantment of Modern* Life, considers the unseen power of objects and how they affect one's body and one's subjectivity.[35] I would argue, and I think Bennett would argue, that if something like a deep map were helpful in seeing new ecological realities, then its ability to create place would be advantageous in a progressive politics for the Anthropocene.

Falling into the traps of Debord's spectacle might well be avoided by a seemingly simple move, which is to have already distilled one's motivations for making the deep map in the first place. If the purpose of the deep map is to answer a historical or theoretical question, for example, then the decisions made while building the deep map should necessarily align with those aims. The specialty of the deep map is its promise to integrate data about a place in a way that takes the user beyond the capacity of his senses. To do so requires an interface of one kind or another through which extrasensorial information is accessed, and this means visual representation. Where there is visual representation, there is the lurking trap of the spectacle because there is the eminent danger of creating a new surface reality, eliding the actual place from the story.

Notes

1 Ben Anderson and Colin McFarlane, "Assemblage and Geography," *Area* 43, no. 2 (2011): 124–7.

2 Jane Bennett, *The Enchantment of Modern Life: Attachments, Crossings, and Ethics* (Princeton: Princeton University Press, 2001), 58.

3 Quotation paraphrased from: Melissa L. Caldwell, "Digestive Politics in Russia: Feeling the Sensorium beyond the Palate," *Food & Foodways* 22, nos. 1–2 (2014): 112–35.

4 Matthew W. Wilson and Mark Graham, "Guest Editorial: Situating Neogeography," *Environment and Planning A* 45 (2013): 3–9; Alberto Cairo, *The Functional Art: An Introduction to Information Graphics and Visualization* (Berkeley: New Riders, 2013); David McCandless, *Information is Beautiful* (London: Collins, 2012).

5 Umberto Eco, *How to Write a Thesis*, trans. Caterina Mongiat Farina and Geoff Farina (Cambridge, MA: The MIT Press, 2015 [1977]); Chapter 4: "The Work Plan and the Index Cards," 107–41.

6 Though from an entirely different context, an excellent example of digital tools operating behind the scenes is: David McClure, "A Hierarchical Cluster of Words Across Narrative Time," *Techne*, July 31, 2017. https://litlab.stanford.edu/ hierarchical-cluster-across-narrative-time/.

7 See also: T.M. Harris, "Deep Mapping and Sensual Immersive Geographies," *The International Encyclopedia of Geography* (2017): 1–13.

8 "Informed speculation" is a phrase I learned from Maria McVarish in her 2019 dissertation at Stanford University's Modern Thought and Literature program. There is currently not available a cite-able version of this work. I have also advocated for similar approaches in geography; see: Nicholas Bauch, "A Scapelore Manifesto: Creative Geographical Practice in a Mythless Age," *GeoHumanities* 1, no. 1 (2015): 103–23.

9 Harriet Hawkins and Elizabeth Straughan, "Introduction: For Geographical Aesthetics," in Harriet Hawkins and Elizabeth Straughan, eds., *Geographical Aesthetics: Imagining Space, Staging Encounters* (Burlington, VT: Ashgate, 2015), 1–17.

10 This is a huge theme in geography and the environmental humanities. For some key examples, see: Sarah Whatmore, *Hybrid Geographies* (London: Sage, 2002); Eduardo Kohn, *How Forests Think: Toward an Anthropology beyond the Human* (Berkeley: University of California Press, 2013).

11 Timothy Morton, "The Mesh," in Stephanie LeMenager, Teresa Shewry, and Ken Hiltner, eds., *Environmental Criticism for the Twenty-First Century* (New York: Routledge, 2011).

12 Richard Misrach and Kate Orff, *Petrochemical America* (New York: Aperture, 2012); Eben Kirksey, *Emergent Ecologies* (Durham: Duke University Press, 2015); Jamie Kruse and Elizabeth Ellsworth, *Geologic City: A Field Guide to the Geoarchitecture of New York* (New York: Smudge Studio, 2011).

13 David Fletcher, "Flood Control Freakology: Los Angeles River Watershed," in Kazys Varnelis, ed., *The Infrastructural City: Networked Ecologies in Los Angeles* (Barcelona: Actar, 2008); Bruno Latour, "Love Your Monsters: Why We Must Care for Our Technologies as We Do Our Children," in Michael Shellenberger and Ted Nordhaus, eds., *Love Your Monsters: Postenvironmentalism and the Anthropocene* (San Francisco: Breakthrough Institute, 2011). Daniel Schneider, *Hybrid Nature: Sewage Treatment and the Contradictions of the Industrial Ecosystem* (Cambridge, MA: The MIT Press, 2011).

14 Graham Harman, *Towards Speculative Realism: Essays and Lectures* (Washington, DC: Zero Books, 2010), 75.

15 William Cronon, ed., *Uncommon Ground: Rethinking the Human Place in Nature* (New York: W.W. Norton, 1996), 57–66; J. Wills, "'Welcome to the Atomic Park': American Nuclear Landscapes and the 'Unnaturally Natural'," *Environment and History* 7, no. 4 (2001): 449–72.

16 J.-D. Dewsbury, "The Deleuze-Guattarian Assemblage: Plastic Habits," *Area* 43, no. 2 (2011): 148–53; quote on p. 150.

17 In some of my recent work I have tried to start this conversation of method. See: Nicholas Bauch, *A Geography of Digestion: Biotechnology and the Kellogg Cereal Enterprise* (Oakland: University of California Press, 2017).

18 Anna Lowenhaupt Tsing, "Empowering Nature, or: Some Gleanings in Bee Culture," in Sylvia Yanagisako and Carol Delaney, eds., *Naturalizing Power: Essays in Feminist Cultural Analysis* (New York: Routledge, 1995).

19 Rona Cran, *Collage in Twentieth-Century Art, Literature, and Culture: Joseph Cornell, William Burroughs, Frank O'Hara, and Bob Dylan* (New York: Routledge, 2016), 5.

20 Jedediah Purdy, *After Nature: A Politics for the Anthropocene* (Cambridge, MA: Harvard University Press, 2015).

21 Weaving as a metaphor for place is from: Robert David Sack, *Homo Geographicus: A Framework for Action, Awareness, and Moral Concern* (Baltimore: Johns Hopkins University Press, 1997).

22 Dewsbury, "The Deleuze-Guattarian Assemblage," 150.

23 For a discussion on mapping and political control, see: Stuart Elden, *The Birth of Territory* (Chicago, IL: The University of Chicago Press, 2013).

24 One relatively early precedent here is: Margaret Wickens Pearce, "Framing the Days: Place and Narrative in Cartography," *Cartography and Geographic Information Science* 35, no. 1 (2008): 17–32. And a more recent articulation of this practice is: Anne Kelly Knowles, Levi Westerveld, and Laura Strom, "Inductive Visualization: A Humanistic Alternative to GIS," *GeoHumanities* 1, no. 2 (2015): 233–65.

25 Jonathan Murdoch, "Towards a Geography of Heterogeneous Associations," *Progress in Human Geography* 21, no. 3 (1997): 321–37.

26 The notion of the technological treadmill as it relates to capitalist agricultural production is found in Keith Dexter, "The Impact of Technology on the Political Economy of Agriculture," *Journal of Agricultural Economics* 28, no. 3 (1977): 211–19. This metaphor taken from agricultural economics fits nicely with an impulse in digital humanities to keep pushing toward new technologies as a means of conducting new research.

27 See also: Matthew W. Wilson, *New Lines: Location-Aware Futures and the Map* (Minneapolis: University of Minnesota Press, 2017).

28 Anne Burdick, Johanna Drucker, Peter Lunenfeld, Todd Presner, and Jeffrey Schnapp, *Digital_Humanities* (Cambridge, MA: The MIT Press, 2012).

29 William A. Koelsch, "Squinting Back at Strabo," *The Geographical Review* 94, no. 4 (2004): 502–18.

30 George H.T. Kimble, "The Inadequacy of the Regional Concept," in John Agnew, David N. Livingstone, and Alisdair Rogers, eds., *Human Geography: An Essential Anthology* (Malden, MA: Blackwell, 1996 [1951]), 511.

31 Kathleen Fitzpatrick, "Beyond Metrics: Community Authorization and Open Peer Review," in Matthew K. Gold, ed., *Debates in the Digital Humanities* (Minneapolis: University of Minnesota Press, 2012). Nina Belojevic and Jentery Sayers, "Peer Review Personas," *The Journal of Electronic Publishing* 17, no. 3 (2014). http://dx.doi. org/10.3998/3336451.0017.304.

32 Guy Debord, *The Society of the Spectacle*, trans. Ken Knabb (Berkeley: Bureau of Public Secrets, 2014 [1967]), 2.

33 Walker Percy, *The Message in the Bottle: How Queer Man Is, How Queer Language Is, and What One Has to Do with the Other* (New York: Farrar, Straus and Giroux, 1975). In chapter 2, Percy writes about the visual "symbolic complex" of the Grand Canyon— how it is almost an entirely visual and non-tactile experience for the vast majority of its visitors.

34 John Lanchester, "You Are the Product," *London Review of Books* 39, no. 16 (2017): 3–10.

35 Jane Bennett, *Vibrant Matter: A Political Ecology of Things* (Durham: Duke University Press, 2010).

4   Spatializing text for deep mapping

May Yuan

The word "mapping" has many meanings. For geographers, mapping is to depict geographic features in graphic forms (e.g., the world map). For mathematicians, mapping is to relate input variables to output variables (e.g., mapping functions). For computer administrators, mapping is to connect hardware parts to software components (e.g., mapping network drives). For neuroscientists, mapping is to associate the structures and functions of the brain and spinal cord (e.g., brain mapping). For cognitive psychologists, mapping is to organize concepts and represent knowledge (e.g., concept mapping or mind mapping). For technologists, mapping is to transform regular objects into artistic expressions of augmented spatial reality (e.g., projection mapping or video mapping).

The idea of deep mapping, while was born out of spatial humanities, encompasses all these ideas of mapping: depicting, relating, connecting, associating, organizing, and transforming. According to Bodenhamer et al. (2015), deep mapping is to engage evidence within its spatiotemporal context and to provide a platform for spatially embedded arguments. One of the main sources of evidence is written documentation that can be used in the diverse means of mapping. Text documentation is the most common form to record observations, document findings, and communicate one's interpretations of observations and findings. Interpretations are driven by the epistemological process to which one subscribes. Deep mapping pushes for the epistemological process that centers on the use of space to frame observations and findings as well as through the spatial framing to contextualize evidence for sense-making. Spatializing text, therefore, enables deep mapping to contextualize texts spatially and temporally and stimulates spatially embedded arguments.

The printing press drives the surge of using text as the main means to disseminate information and inform knowledge. However, these texts are not in a machine-readable form. Intensive archival works progress slowly to identify spatial terms or place markers for graphical transformation, geographic projection, and knowledge contextualization. The proliferation of digital technologies for publishing has diversified the sources and types of documents available for novel text-based studies. Google Inc. in collaboration with libraries around the world launched a historical book project in 2004, and by 2015, they have digitized 25 million volumes of texts in 400 languages from more than 100 countries (Heyman

DOI: 10.4324/9780367743840-4

2015). Google digitizes books and provides full-text indexing that allows word searches, n-gram analytics, and other text-computing algorithms applied to a volume or across the entire collection. Research shows that Google Books approximate the content of a generic major American research library and thus can be considered as a general research collection of books, especially for historical and cultural research (Jones 2011, 77–89; Bohannon 2010, 1600). The Google Books data set, furthermore, inspires a new field of studies: culturomics that applies quantitative methods for cultural research (Michel et al. 2011) and offers unprecedented rich texts "spatializable" for deep mapping. Nevertheless, Google Books corpus inherits biases toward scientific articles, as in most university library collections throughout the 1900s. Analysis of word frequencies, even after resultant normalization, does not necessarily reflect the true cultural popularity of those words in a certain period (Pechenick et al. 2015).

Complimentary to Google Books are the ever-expanding digital newspapers, magazines, journals, blogs, photographs, and videos. These digital media provide rich evidence sourcing for political, social, cultural, and individual accounts of observations and findings, many of which contain time markers by default and hence privilege the production of temporally embedded arguments. For example, the spike of newspaper articles mentioning both "America" and "competition" in the British Newspaper Archive suggested concerns about American competition in response to the harvest of British crops in the early 1880s (Nicholson 2013). Algorithms with temporal notation schemes of dates and times in simple temporal expressions have some success in the temporal resolution of news (Mani and Wilson 2000). In complex temporal references, an event embedded in articles, however, could be difficult to determine when it occurred, observed, or reported, and therefore, temporal ordering of events from multiple digital media remains a hard problem (Navarro-Colorado and Saquete 2016).

In addition to temporal information, deep mapping demands explicit spatial references. Besides challenges faced by resolving events chronologically, toponym resolution accounts for the dynamic and ambiguous nature of place names. Absolute georeferencing systems, such as latitudes and longitudes, while are commonly available on geotagged microblogs (such as tweets or WeChat), few text articles would reference event locations based on geographical coordinates. Geographical Information Retrieval (GIR), emerging as one of the hot topics in Geographical Information Science, aims at developing methods to extract geographical information from text documents (Jones and Purves 2008). While spatializing texts include spatial contexts beyond geographic spaces, GIR advances facilitate identification of event locations from texts. Non-geographic spatial contexts frame evidence based on frequencies or other event properties through visualization. Word clouds, for example, using frequencies and distances among the appearances of words in documents, have shown to be effective in understanding research responses (McNaught and Lam 2010).

This chapter overviews methods that spatialize text and discuss how spatialization fuses evidence in space and time to enrich understanding and interpretation. The overview covers methods for spatializing texts but is unable to include methods for extracting temporal references due to the page limit. Survey papers on methods and applications of temporal information retrieval are available, such as Campos et al. (2015) and Moulahi et al. (2016), for references. Specifically, this chapter reviews the following methods: word clouds, self-organizing maps, and georeferencing. While these methods are applicable to a variety of text analytics applications, the chapter will focus their capabilities on geographic meanings and spatial storytelling. Based on the nature of spatial frameworks, spatialization can proceed semantically or geographically. The following discussion first examines methods spatializing texts in semantic spaces and then methods placing texts in geographic spaces.

Spatializing texts in semantic spaces

Semantic spaces frame texts based on some attributes of words or phrases in texts. From this perspective, spatialization transforms text information into a graphic representation to visualize the content of a document. Word clouds are perhaps the most simplistic and common method to spatialize texts. The popularity of word clouds is promoted by many tools freely available online to generate aesthetic and, at times, artistic word clouds that reveal the frequencies of words, the main topics, and themes in a text. Common tools include Wordle (www.wordle. net/), WordClouds (www.wordclouds.com), WordArt (https://wordart.com/), TagCrowd (http://tagcrowd.com/), and Tagxedo (www.tagxedo.com/). A word cloud algorithm generally includes two functional components. One is to count the occurrence of every unique word in a text. Some word cloud algorithms are case-sensitive; others are not. Most word cloud tools provide options for handling case sensitivity. The second component of the algorithm is to place words in a graphic form. A common approach is to place the most frequent word at the center, place words in the order of frequencies around the most frequent word, and in the end, fill in the gaps with infrequent words. If a word overlaps with any previously-placed word, the word will be moved incrementally along an increasing spiral to eliminate overlaps. For a large text, the placement process can be very time consuming. The efficiency of a word cloud tool depends largely on its placement strategy. Wordle, for example, uses a combination of hierarchical bounding boxes and quadtrees to expedite word placements (Feinberg 2010).

Clement et al. (2009) claimed that "Seeing the manner in which the structure of text makes meaning *in conversation with* narrative alleviates perceived instabilities in the discourse" (p. 8485). They used Wordle to generate word clouds for two comparisons of *The Making of Americans*. One comparison was to a set of reference texts of nineteenth-century novels written by Jane Austen, Charles Dickens, George Eliot, and George Meredith, and the other was between the first and second half of *Making*. The word clouds explicitly showed words that appeared more common in *Making* and less common in the reference texts, and the word "one" surged to be the most prominent highly frequent word in Making compared to the reference texts. The word one also appeared much more frequently in the second half of *Making* than in the first half. They then investigated how the word *one* was used in the text using ProViz, which colored the word *one* according to its part of speech and sized it based on its frequency of a specific word use. In doing so, they identified changes in word usage within a document and across other texts with a new perspective on "the meaning-making processes of the text's composition." Figure 4.1 is an example of spatializing the article *Deep Mapping and the Spatial Humanities* by Bodenhamer et al. (2013). The highly frequent words are spatial, GIS, data, humanities, social, studies, space, time, narratives, and maps. In addition to the words included in the title, the article provokes new thinking in GIS technologies and maps to social studies, place and narratives in space and time.

Word clouds have been used to highlight frequent words in the news, speeches, books, and various texts, including qualitative survey data (McNaught and Lam 2010). From July 11 to September 8, 2016, Gallup polls asked 30,000 Americans what they had heard, read, or seen about 2016 US presidential candidates and created two word clouds available at www.gallup.com/poll/195596/email-dominates-americans-heard-clinton.aspx. The two word clouds were very telling and marked sharp contrasts between the two candidates: email was overriding in the responders' minds about Clinton, while campaign related activities were what the responders recalled about Trump. However, counting words in most situations is not specific enough, and sometimes word clouds can be misleading. Without contextual information that shapes the stories of the presidential election, the dominant "email" in

Figure 4.1 A word cloud representation of Bodenhamer et al. (2013): Deep Mapping and Spatial Humanities.

Clinton's word cloud would be difficult to decipher. Knowing the most frequent words in a text does not help us understand the narratives and arguments in the text. Word clouds divorce individual words from the context and leave readers to figure out how these words relate. Crit words, like "don't" or "against" are often omitted in word clouds. If Bodenhamer et al. (2013) included many statements that GIS technologies don't work for spatial humanities or alike, word clouds that omit "don't" would lead to an opposite sense to the assumed arguments.

Semantic word clouds attempt to preserve semantics so that words semantically more related to each other will be placed closer in a word cloud. Linguistically, two words often appear together in sentences are likely to have a semantic relationship and hence should be placed close to each other in a semantic word cloud. The added criterion for word placement significantly complicates the word cloud formation and makes it an NP (nondeterministic polynomial-time) hard problem (Barth et al. 2014), which means that computational time for the problem can be boundless. Heuristic algorithms, most using force-directed graphs, approximate good-enough layouts for semantic word clouds. Empirical tests showed that Cycle Cover and Star Forest outperformed four other word cloud layout algorithms in preserving semantics (Barth et al. 2014). Semantic word clouds tools with various options are available at http://wordcloud.cs.arizona.edu/. Figure 4.2 is a semantic word cloud showing the 100 most frequent words in Bodenhamer et al. (2013): *Deep Mapping and Spatial Humanities*. Compared to Figure 4.1, the semantic word cloud displays two general ideas in the paper: one that centers on humanities and the other relates to Spatial, GIS, maps, and technologies. Transitioning between the two general ideas are words like geographic, place, location, space, theory, social, history, digital, and deep, which reflect the underlined ideas in the paper to bridge humanities and spatial technologies.

Besides word clouds, Self-Organizing Maps (SOMs) provide means to group documents based on measures of text similarity and project the groups to a two-dimensional space for visualization. In particular, Kaski et al. (1998) developed

Figure 4.2 A semantic word cloud of Bodenhamer et al. (2013): Deep Mapping and Spatial Humanities.

WEBSOM to facilitate the use of SOMs for document analysis. Text similarity usually includes the statistics of words (i.e., word histograms) in each document. Each unique word is considered a variable, and the normalized occurrence of the word in a document is its attribute. Normalization is necessary to avoid biases introduced by the length of a document. The number of unique words among a large collection of documents can be rather big. WEBSOM takes the normalized occurrences of all unique words in consideration to determine similarity among the documents. The more similar word histograms are in two documents, the more similar the two documents are, and therefore, should be placed closer on a SOM.

Kohonen et al. (2000) used WEBSOM to create a SOM based on 6,840,568 patent abstracts in English. On average, each patent contained 132 words. Once they removed words occurred less than 50 times and 1,335 stop words, they reduced the corpus to 43,222 words. Since each word is a variable, the input data to the SOMs were vectors of 43,222 variables, which combined values of occurrences determine where a document should be placed in the SOMs. The spatial representation of a document collection offers keyword search and content addressable search for document retrievals. Keyword searches allow users to seek documents with specified keywords and visualize the distribution of the selected documents on the map to get a sense how common or limited are the keywords to different patent groups. Content addressable searches enable users to identify a subject area of interest and explore approximate topics nearby.

SOMs utilize space to frame and relate texts, which not only empowers a new dimension for document searches but also compresses massive information into a rich map of deeply connected words, documents, topics, and relations. Open-Access WEBSOM examples and figures are available at the permanent link http://urn.fi/urn:nbn:fi:tkk-002573. Many research groups developed similar visual tools to map the semantic spaces of documents have been developed in the early 2000s (Dodge 2005), but most of the tools lacked maintenance plans and ceased to exist. IN-SPIRE™ Visual Document Analysis developed by the Pacific Northwest National Laboratory of the US Department of Energy (http://in-spire.pnnl.gov/index.stm) is a few of fully developed information mapping tools that can automatically analyze a large collection of unstructured texts and position them semantically to elicit common themes and hidden relationships among the documents (Figure 4.3).

Spatializing texts in geographic spaces

To project texts in geographic space, we need some spatial markers to determine the corresponding real-world locations. Spatial markers can be place names, addresses, geographic coordinates, and spatial relations to known locations (such as distance and direction to the Statue of Liberty). The proliferation of smart phones and various mobile devices promotes geotagging text messages and photographs on the go, which provides opportunities to explore activity spaces through narratives. These geotag narratives promote the idea of word clouds in the shape of a geographic region that contextualize the texts geographically;

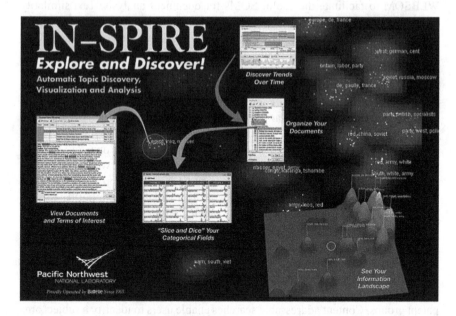

Figure 4.3 IN-SPIRE stands for In Spatial Paradigm for Information Retrieval and Exploration. It maps documents in semantic spaces to use space as the means for information retrieval and analysis.

Source: The image is from http://in-spire.pnnl.gov/about.stm under the courtesy of the Pacific Northwest National Laboratory with permission to free distribution for noncommercial, scientific, and educational purposes.

Buchin et al. (2016) coined geo word clouds to capture not only the frequency of words but also the spatial relevance of words. When each document corresponds to a location, such as a local newspaper, we can place the word cloud for a document to that location. As such, we can build a map of word clouds with geotagged documents to compare spatial differences in news reports, for example. Yahoo! Research developed World Explorer web service that automatically linked geotagged photos from Flickr to locations, extracted the labels with photos, and sized the labels (usually formal or informal place names) according to the number of linked geotagged photos (Ahern et al. 2007). World Explorer, therefore, showed popular locations and common names that Flickr users referred to these locations.

Besides geotagged social media data, Li and Zhou (2017) collected GPS trajectories and activity diary data from 155 individuals between 13 and 19 years of age in four Illinois cities to map the subjects' feelings across each city. The maps are in the form of georeferenced word clouds in which positions of words correspond to the locations where the words were used in narratives. The size of words indicates the frequency of the words appearing in narratives at a location. In doing so, the maps show the spatial distribution of most mentioned feelings across each city. For example, their map shows that clusters of areas where "fun," "excited," "calm," or "relax" are the most prominent words in the subjects' narratives, which might indicate how the individuals took different kinds of activities across each of the cities.

At a large national or international scale, mapping geotagged postings on social media is gaining popularity with various systems developed to collect, parse, and map postings from social media. With geotagged coordinates, social media maps plot locations of postings and display hot spots or clusters of high frequencies of specified keywords. Researchers soon recognized the limitation of keywords in capturing meanings and proposed text mining and sentiment analysis to decipher social media (Pang and Lee 2008), based on which opinion forecasting is possible (Sobkowicz et al. 2012). Spatializing social media postings also progressed from density-based hot-spot mapping to visualize ambient geospatial information (Stefanidis et al. 2013) of activities, relationships, ideas, and concepts (Tsou et al. 2013), pointing to the direction of deep mapping of social media. Noteworthily, Twitter, the most common social media data used for mapping, responds only 1% of randomly selected tweets to any query via their public Application Programming Interface (API), 75% of Twitter users are between the ages of 15 and 25 (Tsou 2015) and only 1% of tweets are geotagged with coordinates (Morstatter et al. 2013). Many researchers include social media or web postings with Internet Protocol (IP) addresses on the device where social media postings were made as proxies of locations to expand locatable text collections. The spatial granularity of locations by IP addresses is mostly at the county or city levels. Examinations of geolocating IP addresses showed that some IP location databases had more than 97% accuracy at the county level and 80% accuracy at the city level (40 km). Other IP address databases resulted at 80% accuracy at the county level and poorly 20% accuracy at the city level (Komosny et al. 2017). The high level of spatial uncertainty deems local mapping inappropriate.

Apparently, georeferencing is straightforward when the locations of texts are readily available with good GPS coordinates. However, location is complicated. Spatial uncertainty is inevitable in any location data and varies with geolocating methods and location itself. The precision and accuracy of GPS coordinates depend not only on the GPS devices but also on several environmental factors (such as atmospheric deflection, structure blockage, multipath effects, etc.) that influence signal reception as well as the availability of axillary infrastructure (WAAS, Wifi, cell towers, etc.) that helps improve location calculation. IP addresses, while spatially coarse, are likely to be the second best option for location inference. When neither GPS coordinates nor IP addresses are available, Natural Language Processing (NLP) offers the third alternative to determine locations based on multiple spatial indicators on meta data and the body of text. NLP may be applied to user profiles (including location field, web sites, time zone, UTC24-Offset) and their social networks to approximate locations or mine actual texts for place names (Schulz et al. 2013). A place name identified from a location field is likely indeed a place name. However, determination of a place name in a text can be very challenging.

There are three main sources of ambiguity in language-based georeferenced (Yuan et al. 2015). Semantically, place names may mingle with person names, title names, feature names, organization names, event names, and other proper nouns in a text. In some situations, place names may be used as metonymies to convey the concept of authority, for example, not to reference a location, such as Washington, the Supreme Court or the White House. Geographically, many places share the same name, or a place name may be referenced to an outdated land mark or with undetermined boundaries. Historically, a place may change names, geographic extents, or representative locations. The entire procedure of matching place names in a text to their correct real-world spatial referents is commonly referred to as toponym resolution, which could involve complicated computation and logical procedures to reach a satisfactory result (Leidner 2008). Standard approaches to toponym resolution are still up for debate. Figure 4.4 represents a general workflow to determine place names in a text. Once places are determined, statements in the text can be mapped geographically, transcending the text to spatial narratives.

Before toponym resolution, language-based georeferenced starts with common NLP to parse text and tokenize parts of speech from which to identify named entities in text, also known as Named Entity Recognition (NER) or Named Entity Classification (NEC), to differentiate people, objects, places, and time markers, for example. Many entities have compound names (e.g., John Doe) or referenced with compound nouns (e.g., the State of Texas), and hence, grouping proper nouns is necessary to catch these compound names. Various text analytics strategies for NER or NEC apply statistical (such as Bayesian) or computational (such as Neural Network) to determine place names based on structural characteristics of chucks and surrounding words in a sentence. The best NER strategy varies with genres and writing styles. Each document collection needs a tailored NER model to achieve desired performance. A subset of randomly selected documents from a collection provides the data to build a NER model, which is then tested and calibrated on another set of randomly selected documents from the collection. Toponym resolution needs to determine if a named entity is a place name and simultaneously functions as a place. Spatial propositions are important indicators along with others, such as terms with state abbreviations, to assure correctly resolved place names.

The NER model building is an iterative calibration process. Important steps include (1) grouping consecutive proper nouns and (2) excluding named entities in non-spatial phrases and stop lists of common names and temporal expressions. When deemed satisfactory, the NER model is then applied to the entire document collection to extract place names. Once a place name is determined, a matching process looks for the place name in the designated digital Gazetteers. Some place names will have multiple matches. For example, there are over 90 cities named "Pleasant Grove" in the US. Determination of which is the right match requires understanding the proper geographic context and assumptions. Miami mentioned in a local newspaper, *Tulsa World* is likely referenced to Miami, Oklahoma than Miami, Florida. A journal noted Paris a 2-hour drive from Dallas is likely to be Paris, Texas, instead of Paris, France. Common parameters used in toponym resolution models to determine the most likely matched spatial referent include

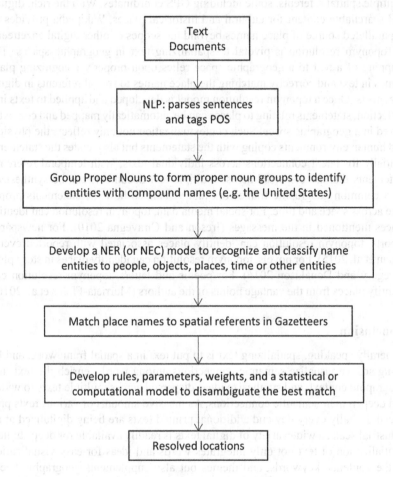

Figure 4.4 A general workflow for toponym resolution in text.

population, distance to previous matched places in the document, containment in the same larger administrative unit (such as in the same state) as other matched places in the document, reachability from the previously matched place if the specific travel means is given, and other geographic measures.

Commonly used Gazetteers or references for place names in the United States (US) or worldwide include US populated places, Geographic Names Information Systems (GNIS), Geonames, OpenStreetMap, Google Map, Alexandria Digital Gazetteers, Wikipedia, and Wikimapia. Many place names relate to rivers, mountains, or some geographic features. US Geological Survey's National Hydrological data and other data on the National Map can be very useful to georeferencing. Wikipedia contains rich information about places, including cities, battlefields, cultural monuments, sports facilities, and so on, and many place entries consist of multiple spatial referents, some including GPS coordinates. With the rich, digital, and searchable content for cultural and historical places, Wikipedia provides an unparalleled source of place names beyond the scopes of other digital gazetteers.

Toponym resolution is pivotal to spatializing text in geographic spaces. The mapping of a text to a geographic space relies upon properly recognizing place names in text and correctly matching the place names to spatial referents in digital gazetteers. Once a toponym resolution model is developed and applied to texts in a collection, statements relating to places can be automatically mapped and contextualized in a geographic space. Such contextualization not only reflects the physical and human environments coping with the statements but also relates the statements spatially for latent connections across individual texts. With temporal referents, statements that are dispersed across multiple text documents can be synthesized on a common geographic framework and visualized how the statements propagate across space and time. For social media data, toponym resolution can identify places mentioned in the messages (Ireson and Ciravegna 2010). For newspaper reports, toponym resolution can identify places associated with reported events (Yuan et al. 2015). For novels, toponym resolution can identify places in story plots (Gregory and Donaldson 2016). For poems and letters, toponym resolution can identify places from the vantage points of the authors (Murrieta-Flores et al. 2016).

Conclusion

Generally speaking, spatializing text is to put text in a spatial framework, and by doing so, we visualize semantic connections among words, enrich the text in a geographic context, and synthesize ideas dispersed across multiple texts to widen and deepen their semantic connections. As massive amounts of various texts produced digitally every day and additional printed texts are being digitalized at an industrial scale, a wide variety of digital texts is readily available for deep studies. Spatialization of text not only integrates words and ideas for easy visualization of the contents, keywords, and themes but also supplements geographic backgrounds to the arguments (or narratives) and elicit their spatial relationships that may remain unnoticed without locating them on a map. The basic idea of spatialization of text for deep mapping grounds individual texts in a massive document collection to provide insights through deep integration of the mass.

The chapter overviews spatialization of text in semantic spaces and geographic spaces. Like the ideas of mapping proliferating in many disciplines, such as geography, mathematics, computer science, neuroscience, psychology, and information technology, deep mapping through spatializing text transforms linear textual descriptions to a two-dimensional spatial framework. Word clouds and SOM are two common methods for text spatialization in semantic spaces. Simple word clouds display words based on their occurrences in the document. The positions of words on a word cloud may be simply random, anchored around the most frequent word and separations of words to the most frequent word in the text, or geographic locations. When a word cloud grounds words in geographic locations, the word cloud shows how different words are used in or to describe different places.

This allows places organically arise as social constructs from the bottom up and with undetermined boundaries, in contrast to administrative places delineated legally or by authority. Mapping these places can reflect human–environment relationships from the mass.

SOMs offer a spatial framework that positions documents based on similarity. There are many similarity measures available to compare documents. A common method is to identify keywords, count frequencies of these keywords in each document, and compare the frequency distributions of keywords (i.e., keyword histograms) across all documents. Documents with similar keyword histograms are deemed to be more similar semantically and will be positioned closer on a SOM. In doing so, spatial relations (or distances) between documents on a SOM reflect their semantic differences. SOMs, therefore, elicit the hidden relations among documents and highlight general themes of document clusters through common, frequent keywords.

Spatializing text in a geographic space requires toponym resolution which identifies place names in the text and determines the spatial referent of the place name. Toponym resolution is challenging because many places have the same name, places can change names over time, and place names may be the same as person's names or be used for purposes other than locations in the text. NLP tools can parse a text and tokenize words into parts of speech. Because many place names are compound names, consecutive proper nouns form proper noun groups, and spatial indicators (e.g., spatial prepositions) can be used to determine which proper noun group may be a place name candidate. Predefined lists of stop words can help filter place names from a candidate list. Once determined, place name candidates proceed to find matches in digital gazetteers. For place names with multiple matches, assumptions impose parameters, weights, and rules in developing a statistical or computation model to determine the best match and, hence, the spatial referent for the place name. Spatializing text is to map these identified place names based on their spatial referent (e.g., geographic coordinates or a digital object in a spatial database) and associate statements with these places. Due to many empirical assumptions and heuristic procedures involved, toponym resolution can be erroneous and, in most cases, needs to tailor procedures and modeling to a specific corpus of documents.

There are many other text spatialization methods, such as concept mapping (Jackson and Trochim 2002) and visualizing information spaces (Fabrikant and Buttenfield 2001). Information visualization and visual analytics are vibrant fields of research, and new methods are being developed at a rapid pace along with the unstoppable deluge of digital information. Spatializing text to support deep mapping, however, goes beyond text visualization. The need to identify and locate ideas and activities documented in texts challenges the fundamental, yet complicated, notions of location and place. Referencing geographic features (e.g., shops, rivers, or neighborhoods) to a location of points, lines, or areas is subject to the assumption that their locations can be precisely determined based on geocoding addresses, GPS coordinates, or name matching in existing spatial data or gazetteers. The assumption is vulnerable considering the unavoidable spatial uncertainty embedded in existing data and algorithms, and more importantly, the multilayered concept about the long-standing problem on "where" and "which" of geographic features with undetermined boundaries (Burrough and Frank 1996). Mapping, deep or not, enforces a location in a space, while text descriptions allow flexible, fluid, and fuzzy associations of features with locations. Deep mapping of ideas and opinions from social media exerts humanity on location and place recognition that leverages the mass to break away from the rigid constraint of technology-enforced spatial precision, such that a mountain is no longer determined by where the cartographer positioned the label on a map but expands across where human experiences prevail through the frequency or semantics of geotagged postings. The desire to spatialize text gives rise to ontological inquiries into geographic complexity and dynamics that complicate the thinking to discern the location, place, and geographic context over time. Furthermore, spatializing text drives new thinking in transforming, projecting, and relating place-based ideas to uncover embedded spatial narratives over time, which is otherwise buried in the linear narration of stories and histories. While this chapter falls short of a comprehensive review of all methods for text spatialization, the limited coverage here introduces the general approaches, relates text spatialization to deep mapping, and highlights references for further readings.

References

Ahern, S., Naaman, M., Nair, R., and Yang, J.H.I. 2007, June. World explorer: Visualizing aggregate data from unstructured text in geo-referenced collections. In *Proceedings of the 7th ACM/IEEE-CS Joint Conference on Digital Libraries*, 1–10.

Barth, L., Fabrikant, S.I., Kobourov, S.G., Lubiw, A., Nöllenburg, M., Okamoto, Y., Pupyrev, S., Squarcella, C., Ueckerdt, T., and Wolff, A. 2014, March. Semantic word cloud representations: Hardness and approximation algorithms. In *Latin American Symposium on Theoretical Informatics*, Springer, Berlin, Heidelberg, 514–25.

Barth, L., Kobourov, S.G., and Pupyrev, S. 2014, June. Experimental comparison of semantic word clouds. In *International Symposium on Experimental Algorithms*, Springer, Cham, 247–58.

Bodenhamer, David J., John Corrigan, and Trevor M. Harris. 2015. *Deep Maps and Spatial Narratives*. Bloomington, IN: Indiana University Press.

Bodenhamer, David J., Trevor M. Harris, and John Corrigan. 2013. 'Deep Mapping and the Spatial Humanities'. *International Journal of Humanities and Arts Computing* 7 (1–2): 170–5.

Bohannon, John. 2010. 'Google Opens Books to New Cultural Studies'. *Science* 330 (6011): 1600.

Buchin, Kevin, Daan Creemers, Andrea Lazzarotto, Bettina Speckmann, and Jules Wulms. 2016. 'Geo Word Clouds'. *2016 IEEE Pacific Visualization Symposium (PacificVis)*, IEEE Asia Pacific, Singapore, 144–51.

Burrough, P.A. and A. Frank. 1996. *Geographic Objects with Indeterminate Boundaries*. Vol. 2. London: CRC Press.

Campos, Ricardo, Gal Dias, Alpio M. Jorge, and Adam Jatowt. 2015. 'Survey of Temporal Information Retrieval and Related Applications'. *ACM Computing Surveys (CSUR)* 47 (2): 15.

Clement, Tanya, Catherine Plaisant, and Romain Vuillemot. 2009. 'The Story of One: Humanity Scholarship with Visualization and Text Analysis'. *Relation* 10 (1.43): 8485.

Dodge, Martin. 2005. *Information Maps: Tools for Document Exploration*. Centre for Advanced Spatial Analysis (UCL), London.

Fabrikant, Sara Irina, and Barbara P. Buttenfield. 2001. 'Formalizing Semantic Spaces for Information Access'. *Annals of the Association of American Geographers* 91 (2): 263–80.

Feinberg, Jonathan. 2010. 'Wordle'. In *Beautiful Visualization: Looking at Data through the Eyes of Experts*, edited by Julie Steele and Noah Iliinsky, 37–58. Sebastopol, CA: O'Reilly Media, Inc.

Gregory, Ian and Christopher Donaldson. 2016. 'Geographical Text Analysis: Digital Cartographies of Lake District Literature'. *Literary Mapping in the Digital Age*: 67–87.

Heyman, Stephen. 2015. 'Google Books: A Complex and Controversial Experiment'. *New York Times*, 28 October.

Ireson, Neil and Fabio Ciravegna. 2010. 'Toponym Resolution in Social Media'. *The Semantic Web: ISWC 2010*: 370–85.

Jackson, Kristin M. and William M.K. Trochim. 2002. 'Concept Mapping as an Alternative Approach for the Analysis of Open-Ended Survey Responses'. *Organizational Research Methods* 5 (4): 307–36.

Jones, Christopher B. and Ross S. Purves. 2008. 'Geographical Information Retrieval'. *International Journal of Geographical Information Science* 22 (3): 219–28.

Jones, Edgar. 2011. 'Google Books as a General Research Collection'. *Library Resources & Technical Services* 54 (2): 77–89.

Kaski, Samuel, Timo Honkela, Krista Lagus, and Teuvo Kohonen. 1998. 'WEBSOM: Self-Organizing Maps of Document Collections'. *Neurocomputing* 21 (1): 101–17.

Kohonen, Teuvo, Samuel Kaski, Krista Lagus, Jarkko Salojarvi, Jukka Honkela, Vesa Paatero, and Antti Saarela. 2000. 'Self Organization of a Massive Document Collection'. *IEEE Transactions on Neural Networks* 11 (3): 574–85.

Komosny, Dan, Paul Pang, Miralem Mehic, and Miroslav Voznak. 2017. 'Evaluation of Device-Independent Internet Spatial Location'. *ISPRS International Journal of Geo-Information* 6 (6): 155.

Leidner, Jochen L. 2008. *Toponym Resolution in Text: Annotation, Evaluation, and Applications of Spatial Grounding of Place Names*. Boca Raton, FL: Universal Publishers.

Li, Dongying and Xiaolu Zhou. 2017. '"Leave Your Footprints in My Words": A Georeferenced Word-Cloud Approach'. *Environment and Planning A* 49 (3): 489–92.

Mani, I. and Wilson, G. 2000, October. Robust temporal processing of news. In *Proceedings of the 38th Annual Meeting of the Association for Computational Linguistics*, Hong Kong, 69–76.

McNaught, Carmel and Paul Lam. 2010. 'Using Wordle as a Supplementary Research Tool'. *The Qualitative Report* 15 (3): 630.

Michel, Jean-Baptiste, Yuan Kui Shen, Aviva Presser Aiden, Adrian Veres, Matthew K. Gray, Joseph P. Pickett, Dale Hoiberg, Dan Clancy, Peter Norvig, and Jon Orwant. 2011. 'Quantitative Analysis of Culture Using Millions of Digitized Books'. *Science* 331 (6014): 176–82.

Morstatter, F., Pfeffer, J., Liu, H., and Carley, K. 2013, June. Is the sample good enough? comparing data from twitter's streaming api with twitter's firehose. In *Proceedings of the International AAAI Conference on Web and Social Media* (Vol. 7, No. 1). The AAAI Press, Palo Alto, CA, USA.

Moulahi, Bilel, Lynda Tamine, and Sadok Ben Yahia. 2016. 'When Time Meets Information Retrieval: Past Proposals, Current Plans, and Future Trends'. *Journal of Information Science* 42 (6): 725–47.

Murrieta-Flores, Patricia, Christopher Donaldson, and Ian Gregory. 2016. 'Distant Readings of the Geographies in Text Corpora: Mapping Norman Nicholson's Poems and Letters'. *CVCE Workshop Proceedings*, Luxembourg.

Navarro-Colorado, Borja and Estela Saquete. 2016. 'Cross-Document Event Ordering through Temporal, Lexical and Distributional Knowledge'. *Knowledge-Based Systems* 110: 244–54.

Nicholson, Bob. 2013. 'The Digital Turn: Exploring the Methodological Possibilities of Digital Newspaper Archives'. *Media History* 19 (1): 59–73.

Pang, Bo and Lillian Lee. 2008. 'Opinion Mining and Sentiment Analysis'. *Foundations and Trends® in Information Retrieval* 2 (1–2): 1–135.

Pechenick, Eitan Adam, Christopher M. Danforth, and Peter Sheridan Dodds. 2015. 'Characterizing the Google Books Corpus: Strong Limits to Inferences of Socio-Cultural and Linguistic Evolution'. *PLoS One* 10 (10): e0137041.

Schulz, A., Hadjakos, A., Paulheim, H., Nachtwey, J., and Mühlhäuser, M. 2013, June. A multi-indicator approach for geolocalization of tweets. In *Proceedings of the International AAAI Conference on Web and Social Media* (Vol. 7, No. 1). The AAAI Press, Palo Alto, California, USA.

Sobkowicz, Pawel, Michael Kaschesky, and Guillaume Bouchard. 2012. 'Opinion Mining in Social Media: Modeling, Simulating, and Forecasting Political Opinions in the Web'. *Government Information Quarterly* 29 (4): 470–9.

Stefanidis, Anthony, Andrew Crooks, and Jacek Radzik owski. 2013. 'Harvesting Ambient Geospatial Information from Social Media Feeds'. *GeoJournal* 78 (2): 319–38.

Tsou, Ming-Hsiang. 2015. 'Research Challenges and Opportunities in Mapping Social Media and Big Data'. *Cartography and Geographic Information Science* 42 (sup. 1): 70–4.

Tsou, Ming-Hsiang, Jiue-An Yang, Daniel Lusher, Su Han, Brian Spitzberg, Jean Mark Gawron, Dipak Gupta, and Li An. 2013. 'Mapping Social Activities and Concepts with Social Media (Twitter) and Web Search Engines (Yahoo and Bing): A Case Study in 2012 US Presidential Election'. *Cartography and Geographic Information Science* 40 (4): 337–48.

Yuan, M., J. McIntosh, and G. DeLozier. 2015. 'GIS as a Narrative Generation Platform'. *Deep Maps and Spatial Narratives: The Spatial Humanities*, edited by D.J. Bodenhamer, J. Corrigan, and T.M. Harris, 176–202. Bloomington, IN: Indiana University Press.

5 Representational issues in deep mapping

Peeling the "poetic and positivistic" from the western geosophical onion

Charles Travis

A major representational issue in deep mapping is how to balance the phenomenological poetry of a sense or personality of place with the positivistic grids of Euclidian and cartographic space. Tensions between these "subjective" and "objective" perspectives have raised representational issues in the arts, humanities and sciences for centuries. The eighteenth-century Italian political philosopher and historian Giambattisto Vico, critiqued his contemporary René Descartes' cartesian methods for studying "civic life," stating that trying "to import the geometric method into practical life" was like "going mad by means of reason . . . as though desire, temerity, occasion, fortune did not rule in human affairs."[1] The geographer Anne Buttimer's phenomenological approach to mapping recognizes the inner contingencies of human agency and observes that our "behavior in space and time [is like] the surface movements of icebergs, whose depths we can sense only vaguely."[2] Thus, one of the main representational issues in deep mapping is how to balance the positivistic indices of latitude and longitude and degrees, minutes and seconds gridding the Cartesian perspective, with the "off-grid" contingencies of the human condition, or what Samuel Beckett states are impressions of "all that inner space one never sees, the brain and heart and other caverns where thought and feeling dance their Sabbath."[3] Trevor Harris states that the intellectual roots of deep maps lie in "eighteenth-century antiquarian approaches to geography, history, people, culture, and place."[4] It is also possible to trace the lineage "deeper" to the practices of the ancient Greeks, medieval cosmographers and early modern cartographers. The American geographer John Kirkland Wright minted the term *geosophy* 1947 to describe the study of geographical knowledge, which he held was analogous to the practice of historiography in history. Distinguishing between cartographical and historical *geosophy*, Wright stated that in regards to the latter:

> The data with which it deals fall within the scope of each and every one of the natural sciences, the social studies, and the humanities. Its conceptions range from the purely personal, subjective impressions of a farmer or a hunter, to those gained by rigorous mathematical calculations and highly refined statistical correlations, and find expression not only in scientific forms but throughout literature and art.[5]

DOI: 10.4324/9780367743840-5

The Greek geographer Strabo (ca. 63 BCE–24 AD), one of the first in the West to have formally practiced *geosophy*, observed that humans come to know the nature and content of the world "through perception and experience alike," which finds confluence in theory, imagination and action.[6] In the seventeenth century Vico coined the term *poetic geography* to describe mapping practices engaged by the ancient Greeks, which continued through to the medieval ages, through which a culture's belief systems, myths and aspirations were projected upon the maps of the landscapes being depicted.[7] *Poetic geography* embellished maps with the world views of ancient Greeks, and Romans depicting battles, sieges, crusades, pilgrimages, spouting sea monsters, and the discovery of new territories. Contemporarily, the *geosophical* perspective of Geographical Information Science (GIScience) on which modern mapmaking relies, largely adheres to mathematical principles established during the Classical, Renaissance, and Enlightenment periods of Western history. This traditional Cartesian perspective adopts an "objective" separation between the perceiver (subject) and the perceived (object). In contrast, (to borrow from the Czech author Milan Kundera) within a phenomenological perspective:

> Man and the world are bound together like the snail to its shell: the world is part of man, it is his dimension, and as the world changes, existence (*in-der-Welt-sein*) changes as well.[8]

Humanistic geographer Buttimer explains: "world to phenomenologist is the context within which consciousness is revealed. It is not -a mere world of facts and affairs, but . . . a world of values, a world of goods, a practical world. It is anchored in a past and directed towards a future."[9] In many ways, language and perceptions "bracketed" by literature, historical, cultural, and cartographical documents (as well as discursive and visual forms of social and broadcast media) reveal multiple slices of idiosyncratic *lebenswelten* constellating across time and space, forming and reforming into unique and contingent "senses of place." Buttimer suggests:

> if people were to grow more attuned to the dynamics and poetics of space and time, and the meaning of milieu in life experience, one could literally speak of the . . . personality of place which would emerge from shared human experience and the time-space rhythms deliberately chosen to facilitate such experiences.[10]

Twentieth-century spatial and cultural turns in the humanities and science have infused a *phenomenological* impulse into GIScience theory, techniques and technology, both countering and complementing the Cartesian perspective, to facilitate the emergence of a humanities-based GIS (HumGIS) deep mapping *geosophical* perspective. One of the key concerns of HumGIS deep mapping is how geospatial technologies can be used to mine, manage, manipulate, chart, visualize, and analyze the subjective geographies embedded within literary, historical, and cultural texts.[11] Attempting to balance the "poetic and the positivistic" through the calibration of "hard" empirical data with "soft" literary, cultural and artistic sources in

HumGIS may point a way to creating symbolic, impressionistic, and quantitative visualizations and analysis of texts, pieces of historical cartography, and other associated corpora.[12] HumGIS deep mapping is arguably a fruit of humanistic geography's late twentieth-century engagement with literature, cinematic, phenomenology, landscape, and "sense of place" studies. Such practices coincide with

> the narrative turn in geography, and those fields with a pronounced geographical imagination, has brought with it a renewed focus on chronology and dating, as thick rather than thin description.[13]

Recently, HumGIS deep mapping techniques employing pastiche, collage, bricolage, and Da-Daist techniques have engaged the kaleidoscopic remediations of urban place that are diffracted through social media platforms and mobile computing and communication devices.[14] Such remediations of *geosophical* perspectives echo Walter Benjamin's observation in *The Arcades Project* (1927–1940) on the changes in *fin de siècle* perceptions of space, place, and time shaped by the invention of the *stereoscope*.[15] Quoting Rudolf Borchardt's *Epilegomena zu Dante*, Benjamin contended the device held the potential to "educate the image-making medium within us, raising it to a *stereoscopic* and dimensional seeing into the depths of historical shadows."[16] Benjamin argued that the *stereoscope* used in conjunction with the gaze of the *flâneur* provided a new means to remap Paris' interpenetrating historical, cultural, and "platial" dimensions. This chapter considers a HumGIS deep map *stereoscopic* juxtaposition of "positivistic" and "poetic" representations of the 1916 Easter Rising in Dublin. The first one focuses on the visualization of archival data on civilian deaths that occurred during the conflict, collected by the Glasnevin Cemetery Trust. The second draws on James Joyce's impressions of the urban atmosphere of the event in *Ulysses*. Though set in 1904, the novel was published in 1922, and contains a Joycean phantasmagoric allusion to the Rising in its *Circe* episode.

Deep mapping: balancing the poetic and positivistic

In the Republic of Ireland, known as the Silicon Valley of Europe, annual observations of the Easter 1916 Rising, and celebration of Bloomsday (June 16) the date in 1904 on which James Joyce set his novel *Ulysses* (1922) are celebrated by postings on *Flickr, Twitter, Instagram, YouTube,* and other social media platforms. Advances in cloud computing now allow the fruit of the Internet of Things, such as digitized archival documents (maps and texts) and social media generated "Big Data," to be processed and integrated remotely on desktop and tablet computers. The synthesis of such data streams can now supply the "deep content" required to analyze patterns and trends about the "many."[17] As many of the 2016 Easter Rising commemoration, and Bloomsday 2014, 2015, and 2018 posts were created on GPS-enabled smartphones, it is possible to harvest the latitude and longitude locations where the text, images, and video they contain were posted. Through creative uses of geospatial technology, these posts can be *respatialized* within a digital time–space fabric woven from the threads of a variety of historical, cultural and literary media, texts, and cartographical documents. Contextualizing posts in a HumGIS deep mapping model contributes to visualizing, in a manner reminiscent to the film *Rashomon* (1950) how the spaces of the 1916 Rising and *Ulysses* intertwine at various cartographical and phenomenological scales to create a variety of senses of place that keep regenerating the human and historical geographies of Dublin.

Quantifying the birth of a terrible beauty

On Easter Monday, April 24, 1916, the occupation of the General Post Office (GPO) in Dublin, Ireland signaled the intentions of a small group of republican and socialist militants to deliver a final proclamation against British Rule in Ireland. The occupation led to an urban military conflict between the Irish rebels and British Army forces which left the city center of Dublin in ruins. The "Rising" was put down by April 30th, and though unpopular among the majority of Dubliners, only achieved totemic status in the Irish political imagination after the rebel leaders were summarily executed by the British in Kilmainham Jail, as depicted heroically in the William Butler Yeats poem *Easter 1916*.[18] In the words of Yeats, the iconic failure of the 1916 Easter Rising in Dublin signaled the birth of an Irish nation. The conflict collapsed civic life in Dublin, but foreshadowed the partitioning of Northern Ireland in 1921, the Irish War of Independence 1919–1921, the formation of the southern 26 county Saorstát Éireann (Irish Free State) in 1922 (the Republic of Ireland was declared in 1949), and ensuing civil war (*cogadh na gCarad*, "war of the friends, relatives") between "Treatyites" and Republicans during 1922–1923. Elements of twentieth-century Irish nationalist historiography created icons and cults of personality out of the Gaelic nationalist Padraig Pearse, socialist labor leader James Connolly respective leaders of approximately 1200 Irish Volunteers and members of the Irish Citizen Army. As illustrated by the valedictory lines of *Easter 1916*, annual Rising commemorations over the course of the twentieth century valorized the protagonists who were executed, and thus transformed by their British executioners into martyrs for the cause of Irish nationhood on an island partitioned into a republican south and unionist north. This contributed to official state narratives that elided the participation of other historic voices and actors. Additionally, and perhaps most troubling, such narratives contributed to the myth of blood sacrifice as a means to achieve a free and united Ireland, providing one rationale, despite the project for nationalist Civil Rights in Northern Ireland for the recrudescence of political violence on the island during the latter half of the twentieth century, including contemporary dissident republican stirrings (see Figure 5.1).

However, the quotidian lives of early-twentieth-century Dubliners celebrated in James Joyce's *Ulysses* (1922)—ordinary people—who were caught between the violent forces of empire and rebellion have only recently been considered-and with good cause. The Easter Rising resulted in at least 485 deaths, according

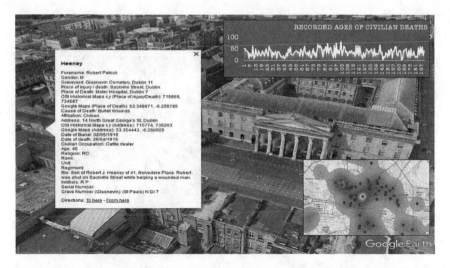

Figure 5.1 Google Earth Mapping of GPO Dublin, Ireland with location of the death of Robert Patrick Heeney, a civilian killed while trying help a wounded man.

Source: Charles Travis, courtesy of Hanna Smyth, the Glasnevin Trust, Google Earth and Esri.

to the Glasnevin Trust. Of those killed, 260 people (about 54%) were civilians.[19] Such as Robert Patrick Heeney a 40-year-old Cattle Dealer, who was shot dead while trying to help the wounded. His death on Sackville Street (now O'Connell Street) took place yards from ground-zero of the Rising—the General Post Office occupied by Pearse, Connolly and their rebel forces. The signing of the Good Friday Agreement in 1998, receding political violence and the presence of a halting but persistent Northern Peace Process has instigated the mapping of the human experiences, agencies and identities of the ordinary lives caught up in the conflict, in contrast to the hagiographic mapping of the events of 1916. In 2016, at the behest of the Irish Dáil (parliament) commemoration committee, a minute's silence was held explicitly in remembrance of all those who died during Easter Week 1916 whether a rebel, in the British Army, the Royal Irish Constabulary (Dublin Metropolitan Police) or a civilian. Accordingly, for the first time, civilian deaths caused by the Rising were publicly and politically acknowledged.[20] The visualization in Figure 5.2 represents the locations of where the lives of combatants and innocent civilians caught up in the urban conflict of the 1916 Rising. One of the main representational issues is to recognize that the locations of the deaths are just a starting point from which to plot a lifepath back to the homes of the deceased to illustrate how near or far they lived from the places they were killed. We could imagine their collective *lifepaths* reaching in to the depths of history and in places intersecting like the roots or branches of a tree, or like the dendritic tendrils of the human brain which carry our memories, hopes, and fears.[21]

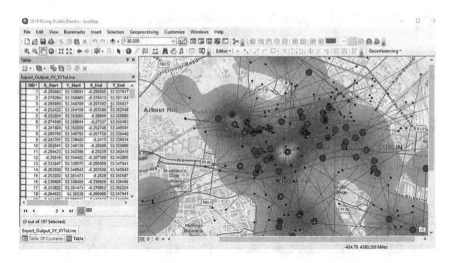

Figure 5.2 Network map connecting location of civilian deaths (white and large circles) to locations of their homes (black dots) by latitude and longitude.

Source: Charles Travis, courtesy of Hanna Smyth, the Glasnevin Trust and Esri.

The locations of these deaths were contextualized by a 2016 *Century Ireland* "Twitter" narrative of the occupation of the General Post Office by the rebels, which allowed individuals to insert their own social media responses to the narrative. The posts illustrated in Figure 5.3 cluster around the 2016 Dublin Rising Commemoration Parade route along O'Connell Street (Sackville Street at the time of the Rising).

Figures 5.2 and 5.3 also illustrate that emerging human centric GIS "knowledge systems" both embrace and transcend Cartesian based geospatial technologies perspectives which frame the relation between "place" and historical events as a skeletal "geometry with names."[22] Indeed, GIScience urban "Platial" models are beginning to integrate social media data, with Big Data streams to facilitate a significant conceptual pattern shift from the classical "layer-cake view of the world" to a digital "networked cupcakes view of the world." Relevant to the evolution of such Hum GIS deep mapping practices, is Jaime Lerner's (2014) theory of "urban acupuncture" in which a city is viewed as a living organism possessing specific "neural" target points that can be targeted and engaged to reenergize its corpus.[23] Connecting the "dots" of these target points, reveals what Seamus Deane defines as a "constellation":

> a previously unrecognized structure or network of relations that was always there, like the unconscious, and appears to us, like it, in articulated images, laden with the weight of the past and yet haloed in the light of discovery and recognition.[24]

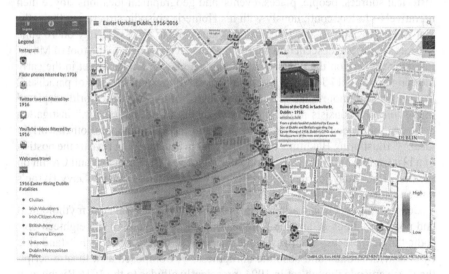

Figure 5.3 2016 Commemoration Social Media Posts juxtaposed with Ground Zero (Black Ellipses) of 1916 Rising Deaths proximate to the General Post Office.

Such an urban acupuncture constellation was revealed by a social media mapping of recent 2014, 2015, and 2018 Bloomsday celebrations in Dublin.

Qualifying the ruin of all space and time

Inaugurated on June 16, 1954 by Irish writers Patrick Kavanagh, Flann O'Brien, and Anthony Cronin, the first official Bloomsday celebration of Joyce's *Ulysses* resulted in an ill-fated attempt to recreate Stephen Dedalus's journey across Dublin in a horse-drawn carriage. Commencing at the Martello tower in Sandycove, south Dublin, the literary sojourn meandered to a few public houses on its way into the city. Either due to boredom, drink, acrimony, or a combination of the three, the celebration imploded at Davy Byrne's Pub on Duke Street, the site of Leopold Bloom's Gorgonzola cheese lunch in the *Lestrygonians* episode. Data harvested from *Twitter*, *Flickr*, and *YouTube*) in Dublin, Ireland with digital Oydessean and Dantean mappings of Joyce's novel were overlaid on a 1904 edition of *Thom's Directory Map of Dublin* as discursive structures to plot the constelled social media "senses of place" (Figure 5.5) generated by contemporary Joycean pilgrims in the Irish capital on successive Bloomsdays.

In addition to the oft-referenced influence of Homer's *Odyssey*, Joyce also appropriated Dante's *Inferno* as a structuring device for *Ulysses*. The medieval vernacular poem's theological *terza-rima* envisions an Italianate Catholic cosmology encompassing an ecclesiastical globe, cleaved by nine levels of an inverted cone fissure, corkscrewing down to the pit of hell (Figure 5.5). Joyce consulted a paperback edition of the *Inferno* to reference the poem's literary and historical sources, people, places, events, and geographical locations. Joyce then narratively plotted, contiguously with his Homeric episodic structuring the paths that Bloom (as Virgil) and Dedalus (as Dante) create across the city on 16 June 1904 as a symbolic descent down through three levels of hell, to the foot of Mount Purgatory, where on the doorstep of Bloom's house on Eccles Street in the early morning hours of 17 June, the pair gain a vision of the constellation of paradise.[25] Mikhail Bakhtin notes that "the temporal logic" of the "vertical world" depicted in the *Inferno* "consists in the sheer simultaneity of all that occurs," instigating a "struggle between living historical time and the extra temporal other-worldly ideal."[26] Illustrating Bakhtin's concept a social media mapping reveals the posting of a black and white June 6, 1954 photograph of writers Kavanagh and Cronin on the first Bloomsday journey juxtaposed with a celebration outside Davy Byrne's pub on Bloomsday 2014 (Figure 5.4).

Deep mapping visualizations can intimate how Joyce layered mythological and poetic readings of an epic European past in *Circe* one of the eighteen episodic "images, ideas, brainwaves and memories" of *Ulysses* which "stand side by side with sudden and absolute abruptness."[27] Composed as a theatre script, the *Circe* episode though set in 1904, presciently alludes to the 1916 Rising as it depicts Bloom and Dedalus' separate paths converging at Bella Cohen's brothel in 'Nighttown'. We can see from the deep map in in Figure 5.6 that the *Circe* episode corresponds to the Middle Level of Hell. In the previous episode *Oxen of the Sun*, both men find themselves in separate company at the National Maternity Hospital on Holles Street. Bloom is visiting Mina Purfoy who is in labor, and Dedalus, on a drinking spree with Trinity College medical students, sets out from

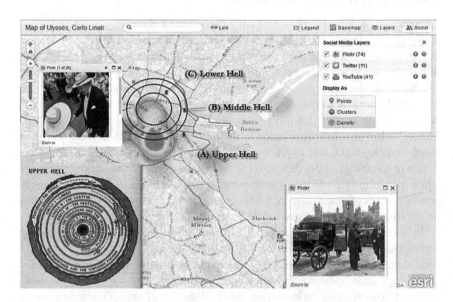

Figure 5.4 Bloomsday "Sense of Place," Dublin 1954 and 2014.

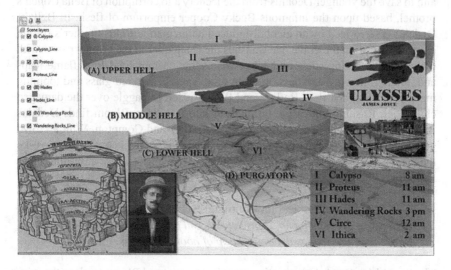

Figure 5.5 Ulysses Deep Map of Dublin.

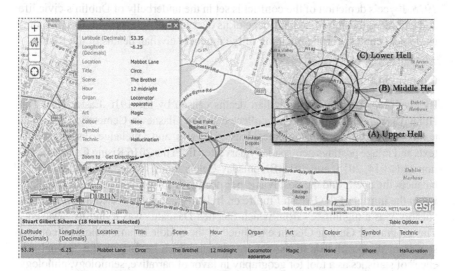

Figure 5.6 Odyssean and Dantean Mapping of Dublin, Ireland.

the hospital on an absinthe-fueled pub crawl before heading to Dublin's north-side brothel district.

Famous throughout Europe from the early nineteenth century, Catholic tradition and constabulary tolerance conspired in the city's "Monto" district to allow whole streets of houses to be used openly as brothels.[28] Bloom, a friend of Dedalus's father, follows the stumbling student through the Amiens Street Railway Station (now Connolly) into the north-inner-city red light quarter with the ironic aim to save the younger Dedalus from the lechery and corruption of Bella Cohen's brothel, based upon the infamous Becky Cooper emporium of flesh. In Bella's, Dedalus is terrorized by an alcohol-induced hallucination of his mother's rotting corpse rising from the grave. He smashes the chandelier in the brothel parlor with his ashplant walking stick, and Joyce writes that "Times livid final flame leaps and, in the following darkness, the ruin of all space, shattered glass and toppling masonry."[29] Stephen flees Bella's, and Bloom is left to haggle over the damaged chandelier. Rushing out of the brothel out into the street, Bloom finds Dedalus in a heated argument with the British soldiers Carr and Compton. With a crowd gathering, Dedalus receives a punch from the soldiers for his exhortations. Like the damned in the *Inferno*, phantasmagorical voices in *Circe* hint at military conflagrations that will raze the city during the 1916 Easter Rising:

> (DISTANT VOICES: Dublin's burning! Dublin's burning! On fire, on fire!) (*Brimstone fires spring up. Dense clouds roll past. Heavy Gatling guns boom. Pandemonium. Troops deploy. Gallop of hoofs. Artillery . . .*).[30]

The Royal Irish Constabulary arrive, people disperse, and Bloom tends to the injuries of the younger man. In contrast to Yeats' ruefully, heroic lamentation in *Easter 1916*, Joyce's depiction of the conflict is set in the underbelly of Dublin's civic life and embraces the quotidian landscape of its everyday people, providing a phenomenological context for the mapping of the innocent civilian deaths of the Rising.

Conclusion

By *stereoscopically* plotting civilian deaths caused by the 1916 Easter Rising in HumGIS with the empirical data collected by the Glasnevin Cemetery Trust, contextualized by the phenomenological lens of *Ulysses*, and a mélange of social media texts and image postings geo-spatially harvested from the 2016 Rising Commemoration and 2014, 2015, and 2018 Bloomsday Celebrations, a deep map model which balances the poetic and the positivistic perspectives can be considered. Joyce (alluding to the impact of the British Empire on Ireland) ruefully reflected in *Ulysses* that "history was a nightmare" from which Stephen Daedalus was trying to "awaken." The geographer Gunnar Olsson exclaimed in the middle of an ontological conversion from the positivistic to the poetic—during which he forsook the "spatial science" of statistics as a tool for geography in favor of narrative, semiology, philology and linguistics—that if Joyce could take such a view on history, then he could let his own pen write that "geography is a prison house from which I am trying to escape."[31] In contrast, John Lewis Gaddis imagines history as a kind of mapping, and links its ancient practice within the archetypal triptych of perceived time (past, present, and future). As a historian, Gaddis notes that cartography and narrative are both practices that attempt to manage infinitely complex subjects by imposing abstract grids and concepts -such as longitude and latitude to frame landscapes or hours and days to denote timescapes. In this manner if the past is conceived as a landscape and historical narrative the way we navigate and represent it, then the recognition of patterns constitutes a primary form of human perception.[32] The HumGIS *geosophical* model presented in this chapter engages in poetic and positivistic pattern recognition to illuminate Harris's view that one of the aims of deep mapping:

> is to shift from a view of humans as entities or data points to an examination of behavior, the material and imaginary worlds, and the relationships that compose notions of a nuanced, non-reductionist, deeply contingent, and scaled conception(s) of place.[33]

Conceiving and incorporating phenomenology, literary, and cinematographic aesthetics into geospatial technology models to represent past and present "senses of place" raises significant representational challenges, and problems to be solved.[34] Michael Goodchild notes that over the past two decades GIS has gradually transformed into a type of media in which "place" perspectives have gained ascendancy.[35] Accordingly, the deep mapping heuristic allows us to integrate the landscapes of time and place theoretically, empirically, statistically and phenomenologically. This approach is reasserting modes of inquiry on human meanings of place and time. In addition to geospatial technology, deep mapping techniques can draw upon a plethora of visual and textual software tools to explore and map the variety of cultural spatial ontologies, languages and landscape perceptions that operate and occupy different regions of the earth, create counterfactual and speculative historical landscapes (past, present, and future) in addition to contributing to public history and geography place-making projects. Deep mapping HumGIS approaches proscribe certain types of hermeneutic, cultural, affective, and subjective representations and analyses, in contrast to simply employing traditional GIScience "paint by numbers" database framings and methodologies. However, as Martin Dodge aptly notes much work on deep mapping "is often misleading in its actual 'map' aspects" because of researchers' lack cartographical knowledge and engagement with cartographic conventions for effective communication.[36] It must be recognized that an understanding and application of *positivistic* traditional cartography methods can act as scaffolding, such as the structure employed by Leonardo da Vinci when lying on his back applying the *poetic* cosmology of *The Last Judgment* on the ceiling of the Sistine Chapel. It is apt to recall that Da Vinci's combined studies in arts and sciences between 1490 and 1495 (particularly hydrology and perspective and proportion) contributed to his masterpieces *The Last Supper* (1498) and the *Mona Lisa* (1503). Equally, the scientist James Lovelock's formulation of the *GAIA Hypothesis* (positing the earth as a self-regulating system) was informed in part by interactions with the writer William Golding, author of the novel *Lord of the Flies* (1954). Such *poetic-positivistic*, mixed method collaborations intimate the potential of deep mapping suggested by Olsson's observation on remote sensing practices:

> for what is that type of mapping at a distance if not a human activity located in the interface between poetry and painting? What is a satellite picture if

not a peephole show, a constellation of signs waiting to be transformed from meaningless indices into meaningful symbols?

Deep mapping considered this way may help illuminate and represent past and present "senses of place" from our human condition's "latent substratum of experience."[37] By revisiting the idea of *geosophy*, harnessing nontraditional geo-data sources and employing a palette of affect and subjective perception, the deep mapping heuristic can inform conventional GIScience and cartographical approaches how to paint with the colors and plot with the vagaries of human experience and behavior through the integration of poetic and positivistic methods.

Notes

1 G. Vico, *De Antiquissima Italorum Sapientia* (*On the Most Ancient Wisdom of the Italians*), trans. J. Taylor (New Haven and London: Yale University Press, 2010 [1710]), 11.
2 A. Buttimer, "Grasping the Dynamism of Lifeworld," *Annals of the Association of American Geographers* 66, no. 2 (1976): 287.
3 Samuel Beckett, *Molloy*, trans. Samuel Beckett and Patrick Bowles (Paris: Olympia, 1955), 10.
4 T.M. Harris, "Deep Geography-Deep Mapping: Spatial Storytelling and a Sense of Place," in David J. Bodenhamer, John Corrigan, and Trevor M. Harris, eds., *Deep Maps and Spatial Narratives* (Bloomington and Indianapolis: Indiana University Press, 2015), 42.
5 J.K. Wright, "Terrae Incognitae: The Place of Imagination in Geography," *Annals of the Association of American Geographers* 37 (1947): 14.
6 Wright, "Terrae Incognitae."
7 J. Gillies, *Shakespeare and the Geography of Difference*, vol. 4. (Cambridge: Cambridge University Press, 1994).
8 M. Kundera, *The Art of the Novel* (London: Faber and Faber, 1988), 35.
9 Buttimer, "Grasping the Dynamism of Lifeworld," 246.
10 Buttimer, "Grasping the Dynamism of Lifeworld," 290.
11 I. Gregory, S. Bushell, and D. Cooper, *Mapping the Lakes: A Literary GIS*, 2013. www.lancaster.ac.uk/mappingthelakes/.
12 D. Bodenhamer, J. Corrigan, and T.M. Harris, *The Spatial Humanities: GIS and the Future of Humanities Scholarship* (Bloomington, IN: Indiana University Press, 2010); Michael Dear, et al., *Geohumanities: Arts History, Text at the Edge of Place* (Abingdon, UK: Routledge, 2011); C. Travis, *Abstract Machine: Humanities GIS* (Redlands, CA: Esri Press, 2015).
13 S. Daniels and P.J. Bartlein, "Charting Time," *Annals of the American Association of Geographers* 107, no. 1 (2017): 28–32.
14 Travis, *Abstract Machine*; D. Cooper, C. Donaldson, and P. Murrieta-Flores, eds., *Literary Mapping in the Digital Age* (Abingdon, UK: Routledge, 2016); J. Stadler, P. Mitchell, and S. Carleton, *Imagined Landscapes: Geovisualizing Australian Spatial Narratives* (Bloomington, IN: Indiana University Press, 2015).
15 W. Benjamin, *The Arcades Project* (Cambridge, MA: Harvard University Press, 1999).
16 R. Borchardt, *Epilegomena zu Dante*, vol. 1 (Berlin: Ernst Rowohlt Verlag, 1923), 56–7.
17 D. Sui and M.F. Goodchild, "The Convergence of GIS and Social Media: Challenges for GIScience," *International Journal of Geographical Information Science* 25, no. 11 (2011): 1737–48.

18 "Easter 1916," in Mary Ruby, ed., *Poetry for Students*, vol. 5 (Gale, 1999), 89–107. Gale eBooks. link.gale.com/apps/doc/CX2691300018/GVRL?u=txshracd2597&sid=GVRL&xid=958dd333 (accessed January 20, 2021).
19 Glasnevin Trust, *1916 Necrology/485*. www.glasnevintrust.ie/visit-glasnevin/news/1916-list/.
20 C. Travis and H. Smyth, "Tell the Story of Irish Public History," *Learn ArcGIS*. Esri Press, 2016. https://learn.arcgis.com/en/projects/tell-the-story-of-irish-public-history/.
21 *1903 Thom's Map of Dublin* image courtesy of Glucksman memorial map library, Trinity College Dublin.
22 G. Olsson, *Abysmal: A Critique of Cartographic Reason* (Chicago, IL: The University of Chicago Press, 2007), 137–8.
23 J. Lerner, *Urban Acupuncture* (Washington, DC: Island Press, 2014).
24 S. Deane, "Walter Benjamin: The Construction of Hell," *Field Day Review* 3, nos. 2–27 (2007): 10.
25 C. Slade, "The Dantean Journey through Dublin," *Modern Language Studies* (1976): 12–21.
26 M.M. Bakhtin, *The Dialogic Imagination: Four Essays*, trans. Michael Holquist, ed. Caryl Emerson and Michael Holquist (London and Austin: University of Texas Press, 1981), 157–8.
27 A. Hauser, *The Social History of Art: Naturalism, Impressionism, The Film Age*, vol. 4 (London: Routledge, 1999), 255.
28 C. Travis "Visual Geo-Literary and Historical Analysis, Tweetflickrtubing, and James Joyce's Ulysses (1922)," *Annals of the Association of American Geographers* (2015). http://dx.doi.org/10.1080/00045608.2015.1054252.
29 J. Joyce, *Ulysses*, ed. Declan Kiberd (London: Penguin, 1992 [1922]), 683.
30 Joyce, *Ulysses*, 694.
31 G. Olsson, "Invisible Maps: A Prospectus Geografiska Annaler," *Series B, Human Geography* 73, no. 1 (1991): Meaning and Modernity: Cultural Geographies of the Invisible and the Concrete, p. 85.
32 J.L. Gaddis, *The Landscape of History: How Historians Map the Past* (Oxford: Oxford University Press, 2002).
33 Harris, *Deep Geography-Deep Mapping*, 18.
34 S. Roche, "Geographic Information Science II: Less Space, More Places in Smart Cities," *Progress in Human Geography* (2015): 1–10.
35 M.F. Goodchild, "Formalizing Place in Geographic Information Systems," in L.M. Burton, S.P. Kemp, M.-C. Leung, S.A. Matthews, and D.T. Takeuchi, eds., *Communities, Neighborhoods, 541 and Health: Expanding the Boundaries of Place* (New York: Springer, 2011), 21–35.
36 M. Dodge, "Cartography I: Mapping Deeply, Mapping the Past," *Progress in Human Geography* 41, no. 1 (2017): 89–98, 9.
37 Dodge, "Cartography I."

6 Indigenous deep mapping

A conceptual and representational analysis of space in Mesoamerica and New Spain

Patricia Murrieta-Flores, Mariana Favila-Vázquez, and Aban Flores-Morán

Deep mapping has emerged as a means of reflecting on and, to a certain extent, contesting hegemonic forms of spatial understanding. The critical analysis of the epistemological roots of technologies such as Geographic Information Systems, and the Western/Cartesian/Euclidean lens that they impose, have been at the core of the arguments to develop a Spatial Information System that could have a strong basis in humanities studies and which should holistically consider the conceptual complexity of space in a variety of cultural contexts. Although there is not yet an embodiment of this idea in the form of "software" or a single technical solution, deep maps can be seen as a step toward acknowledging this complexity. Interestingly, although deep mapping is conceived as a modern methodological approach in the Spatial Humanities, there are forms of deep mapping that have existed for millennia, showcasing how societies have created multimodal statements of their own spatial understandings, history, and cosmogony. This is the case with Mesoamerican codices and colonial "maps" (called *pinturas* or *lienzos*—paintings—by the Spaniards), that would have an oral accompaniment and could be conceived as full ideological, social, symbolic, spatial, and temporal mnemonic expressions of the societies that created them.

With the European encounter of a different worldview and system, these expressions suffered multiple transformations. As thousands of prehispanic codices were destroyed and new conceptions, writing systems, perspectives, and a new religion were introduced, Indigenous colonial painters continued to express themselves through their own traditions, while becoming familiar with the European ones. New colonial maps and codices would emerge, incorporating prehispanic imagery and iconography while adopting, adapting, and continuously reinterpreting Western spatial and pictorial notions. The purpose of this chapter is threefold. First, we aim to delve into the ways that Mesoamerican and colonial native societies conceived spatial and geographic depictions, portraying the complexity of native cosmogony through pictorial expressions that intertwine conceptions and narratives of both real and mythical space–time with those of power, the sacred, and the community. Second, we look to provide a reflection on the syncretic processes that took place, through the analysis of some crucial elements observed in three sixteenth-century colonial spatial representations. Finally, we highlight the idea that, for the majority of societies in the world (possibly with the exception of the

DOI: 10.4324/9780367743840-6

modern), spatial conceptions usually went beyond Euclidean mapping, and created expressions that equate to deep mapping that persist even today. In doing so, we keep advocating for a decolonial approach in the Humanities and Sciences, and the need to delve into the study of past and subaltern societies, with a constant reflection upon the use of technologies that are rooted in Western/modern conceptions.

Introduction

When the Spaniards arrived in the so-called New World's unknown territories in 1519, they encountered not only societies that they had never heard about but also an unfamiliar land and the prospect of wealth that could benefit the Spanish Crown and the new settlers. In the first months after reaching the mainland and starting the journey inland to the famous capital of the Mexica Empire, Mexico-Tenochtitlan, Hernán Cortés, and his soldiers had to be guided by Indigenous people in order to reach their destination. Once Cortés was granted an audience with the *tlatoani* Moctezuma II, he asked for some maps of the east coast so that the conqueror could locate a suitable place to harbor his ships. In his second letter, Cortés describes how Moctezuma ordered to be delivered to Cortés, cotton textiles depicting the southern territory of Veracruz's current state and particularly the Coatzacoalcos region (Cortés 1856). This would become probably the first time a European had in his possession what we assume were Indigenous maps.

After this episode, Cortés requested maps to be provided by noblemen from other regions, at least one more time. By the time he arrived in the Hibueras area, modern Honduras, Cortés had acquired another map that the Tabasco ruler had previously given him. He wanted to reach the province of Chilapa, and when trying to navigate the route to his destination using the Indigenous map, he failed to do so. He got lost for a few days until, finally, local guides helped him. When reading this episode in the fifth letter that Cortés sent to the King of Spain, we could think that the maps were not geographically accurate, consequently explaining why Cortés could not reach Chilapa. But this may be otherwise explained by the fact that Cortés was used to reading nautical charts and maps indicating the location of sites under a specific spatial reference system, and a conception of the world that did not match the one he saw in the *pinturas* (paintings) Moctezuma and others provided him.

Once the Spaniards and their Indigenous allies defeated the Mexica army and conquered Mexico-Tenochtitlan in 1521, a series of negotiations took place to create and organize what would become known as the Viceroyalty of New Spain. To achieve this, it was necessary to amass a large amount of information on the vast new territories under the control of the Spanish Crown. Such an enterprise was primarily led by the Royal Cosmographer Alonso de Santa Cruz from 1553 to 1567, and under the reign of Charles V (Mundy 1996, 12). Later, it would be Juan López de Velasco, the new royal "main chronicler-cosmographer" of King Philip II, who took over the project in 1571 and sent a questionnaire to New Spain so he could compile information on people, geography, resources, and organizations among other topics, along with maps of the recently created provinces. The result of this enormous effort would become known as the *Relaciones Geográficas del siglo XVI* (the Geographic Reports from the sixteenth century) (see Cline 1972; Mundy 1996; Portuondo 2009).

The main objective of the cartographic project associated with the compilation of the *Relaciones Geográficas* was to obtain maps that could later be integrated to become an extensive atlas of New Spain. Once the local authorities answered the questionnaire, these reports were then sent to Spain along with the maps, though the latter were not always included. Juan López de Velasco received 91 maps and the corresponding texts answering the questionnaire according to the *Real Cédula* (Royal Command) issued by Philip II, May 25, 1577 (Robertson 1972). Given the instructions he outlined in the questionnaire,[1] we can infer that López de Velasco might have been profoundly disappointed with the results. While he was expecting measurements and cartography in the form he was used to, what he received instead were spatial representations based on Indigenous tradition, combined with some words in Spanish and mostly Nahuatl (the language of the Mexica Empire also known as Aztec), and sometimes cartographic elements of the European tradition (Mundy 1996, 23–7). Of the 91 maps that came into de Velasco's hands, only 35 were produced following the rules of Western cartography. These representations have been classified as "European" by Robertson (1972), while the remainder had a "Native" or "Mixed" style.

While Cortés was confused when trying to read the document of Mesoamerican tradition received from Indigenous rulers, the Royal Cosmographer was surely able to identify some elements in the images he observed because by that time European conceptions were permeating throughout the Viceroyalty. Nevertheless, his frustration must have been more considerable when he could not integrate them into a familiar coordinate system. As a consequence of this, and other problems with the written information he received from the New World, López de Velasco would resign his administrative position in 1588, without actually concluding the primary objective of his project to provide King Philip II with a comprehensive image of his distant realms in New Spain (Mundy 1996, 9).

The paintings given to Cortés by the local rulers did not survive. Still, thanks to his own chronicles and those of the soldier Bernal Díaz del Castillo, we know that they recognized them, to a certain extent, as cartographic documents with some identifiable representations of the surrounding landscape (Castañeda de la Paz and Oudijk 2011, 87). From these and other historical accounts such as the maps of the *Relaciones Geográficas* (although these take place in the last phase of the sixteenth century), we can infer the existence of prehispanic representations with some cartographic characteristics (León-Portilla and Aguilera 2016, 18–22). To this, we can add the spatial representations present in the prehispanic codices and the scarce evidence found in some scenes depicted in mural paintings, particularly from the Maya cultural area (Estrada-Belli and Hurst 2011).

This evidence allows us to detect a fundamental difference between Western maps and Mesoamerican representations and in which the cosmogonic character of space is represented in this latter tradition. Mountains, rivers, plants, and earth were conceived as living beings, taking a primary role in pictographic narrations as observed in the surviving prehispanic codices, where nature not only intervenes in the narrative but is also intertwined with human actions and destiny. This conception strongly differs from Western-cartographic depictions. Although the European concept of landscape (derived from the German *landschaft*) might have been connected to community, territory, and its environment before the sixteenth century (Fernández-Christlieb 2015, 335), the concept and its development were used under the reigns of Charles V and Phillip II to create graphic records of territory with economic and political intentions. To the contrary, Mesoamerican conceptions of landscape cannot be separated from an intricate weave of historical, symbolic, religious, and ritual connotations. As such, this assemblage of evidence would suggest that spatial Indigenous representations may be better thought of as mnemonic spatial technologies (similar to "deep maps") that could include diverse elements that were recognizable to the Western eye. Nevertheless, these representations included elements that were ambiguous to the European, such as historical and mythical narratives (identifiable by calendrical glyphs); itineraries (represented by footprints); toponymic glyphs; geographic locations; mythical sites that do not necessarily have a geographical correspondence; physical places that are at the same time mythical places; anthropomorphic, landscape and architectural representations, and genealogies linked to the government of a place. Many of these characteristics are also found on later maps created after the Conquest, and they were the elements that might have disoriented the Europeans at first sight, as in the case of López de Velasco.

The indigenous conception of the cosmos

To understand the spatial narratives that we observe in the codices of Indigenous tradition, and later in the maps of the second half of the sixteenth century, it is worth delving into the conception of the world held by the authors of these paintings. The structure of the Mesoamerican cosmogony has intricate qualities that, to date, have not been fully understood. Modern interpretations of the structure of the cosmos in the Indigenous worldview suggest the existence of three interconnected spaces: the sky, the earth's surface, and the underworld. The sky has 13 levels (*chicnauhtopan* in Nahuatl), and the underworld (*chicnauhmictlan* in Nahuatl) has nine levels through which deities could transit (López-Austin and López-Luján 2009, 43).[2] In the middle was the earth's surface, conceived as a living being of reptile characteristics that was floating in the water, named Cipactli by the Mexica.[3] This being had unique characteristics: its back was corrugated, which corresponds to the world's topography; in the center (*axis mundi*) there was a cosmic tree or a sacred mountain connecting these three realms and allowing the transit of deities and nature forces through these plains (Manzanilla 2008, 113; López Austin and López-Luján 2009, 22). This earth monster was inhabited by humans, animals, plants, and natural phenomena through which deities were present, creating an animated and divine world, where the sacred intertwines with the profane.

Prehispanic spatial representations, just like deep maps, were not reduced to the tangible or material, but included and depicted ideological dimensions of space and time through certain elements: (1) toponymic glyphs; (2) genealogies and their relationship with deep conceptions of time; (3) Indigenous governors; (4) individual geographic features; and (5) material culture. These elements were integrated through the Indigenous concept of *altepetl*,[4] which was still present on the maps requested by the Spanish crown in the sixteenth century.

Altepetl: a prehispanic spatial conception

During prehispanic times, ideas about territories, human settlements, and land differed profoundly from Western concepts with which we are more familiar. The Nahua word *altepetl*[5] is a key spatial prehispanic concept translated to Spanish in sixteenth-century dictionaries as "the town, the city, the province, the king" (Molina 2008). This polysemic translation responds to the complex connotations of this concept that refers to a community organized under economic, political, social, and religious principles, which inhabits a specific territory that was not a single continuous extent, but was an integration of individual units distributed through space (Gibson 1964; Lockhart 1992; Florescano 2009; Gutiérrez Mendoza 2012; Fernández-Christlieb 2015). Each *altepetl* was composed of minor units called *calpolli* (borough), featured a temple, and was ruled by a *tlatoani* (governor). At the same time, each *altepetl* could be part of a larger confederation called a *hueyaltepetl* (large altepetl), which was ruled by a *hueytlatoani* (great governor).

The concept of *altepetl* simultaneously refers to the symbolic conjunction of the aquatic realm and the earth realm, in the form of a sacred mountain containing water. This mountain was the *axis mundi*, and was connected to the sea by springs, rivers, and subterranean water. The precious liquid came out of the mountain through caves that were conceived as liminal places and entrances to the earth's interior, where humans could get in touch with the celestial and underworld realms (Broda 1991: 479–80, 2015).

Each human settlement belonged to a geographical *altepetl*, which would generally be a hill or high mountain close to the town where people went and made offerings and performed rituals (Broda 1991). We can find pictographic representations of these mountains/*altepeme* (pl.) in many codices as a bell-shaped figure with a tucked base and voluted lines in the outline of the glyph that refers to the stony surface of the mountain. At times, this surface is filled with a rhomboidal pattern with a circle in the center that would be a representation of the earth monster's skin (Figure 6.1).

The correspondence between each Indigenous town and its *altepetl* was represented through a toponymic glyph; a representation of the name of a community or a specific place (Castañeda de la Paz and Oudijk 2011, 91).[6] To register the toponym, "phonetic transcriptions" were used, that is, the superimposition of figures that graphically do not cohere, as the purpose of each was to provide a sound to create part of a word (Hill Boone 2010, 48–9) (Figure 6.2a). In many cases, one part of this glyph is the representation of the mountain, and the other can make reference to a geographical characteristic close to where the town is settled and could be interpreted as an "abstract landscape representation" (Figure 6.2b) (Castañeda de la Paz and Oudijk 2011, 91). In many other cases, however, the reference of the

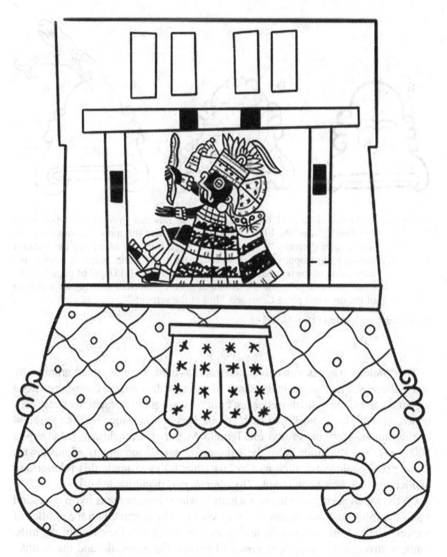

Figure 6.1 Representation of Mount Tlaloc on page 22 of the Codex *Borbonicus*.

Source: Drawing by Aban Flores-Morán.

glyph is not clear-cut, and its presence has to be inferred by its relation to a mythical narrative or even function as a phonetic element (Figure 6.2c).

Each *altepetl* had an associated deity related to the economic activities carried out by the people, and to its mythical and historical origins (Navarrete Linares 2011, 25). These deities were worshiped in a temple that was a representation of the sacred mountain, and consequently, the center of the world, where political and religious power resided (López-Austin and López-Luján 2009, 18).

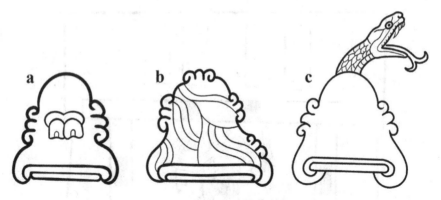

Figure 6.2 (a) Detail of page 11 of the *Matrícula de Tributos* showing the toponym of
Tepetitlan, "near the hill." The two teeth inside the mountain represent the
suffix -tlan ("nearby"). (b) Detail of page 5 of the *Matrícula de Tributos*
showing the toponym of Tezoyucan as a mountain made of stone, the
translation of the name is "place filled with stone." (c) Detail of page 5 of the
Codex *Boturini*, also known as Tira de la Peregrinación, showing the toponym
of the mythical place Coatepec, "hill of the serpent."

Source: Drawing by Aban Flores-Morán.

This analogy was not exclusively conceptual but was represented graphically
and materially. The temple, in the shape of a pyramid, emulated that of the sacred
mountain and some of them were even built on springs to represent the water contained within. Two examples of such "mountains of sustenance" are the Pyramid
of the Sun in Teotihuacan and the Templo Mayor of Tenochtitlan (Manzanilla
2000, 99). The pyramids were the hand-made sacred mountains of the towns.
Some prehispanic cities such as Cholula (Puebla) even made this relationship
explicit, calling their main temple *Tlachihualtepetl* (hand-made hill).

On the top of such pyramids was a temple whose entrance was like a cave, and
when the rulers and priests entered, they could make contact with the entities of
the celestial and the underworld levels, as was the case in the caves of the hills.
Thus, a strong association was created between the mountain and the temple,
since both were conceived as an *axis mundi* and as representations of the *altepetl*.

Another example of this analogy is located in page 14 of the Codex *Borgia*, where
the facade of the structure is represented as the maw of the earth-monster (Ruiz
Medrano 2001, 156), with the same rhomboidal-circle pattern that alludes to its skin
(Figure 6.3). The image is complemented by a character dressed as the deity *Tepeyollotl* who goes to the interior of the temple holding an offering in his hands, where he
will come into contact with the forces of nature. This scene is one of the best representations of the confluence of the temple and the mountain as signs of the *altepetl*.

The *altepetl* was linked with the ruler to such an extent that the representation of one could be exchanged for the other. Thus, in different prehispanic

Figure 6.3 Detail of page 14 of Codex *Borgia* depicting *Tepeyollotl* in front of a temple.

Source: Drawing by Aban Flores-Morán.

representations, when referring to a victory, defeat, conquest, or any other out-standing fact happening in a town, the ruler's image represented the city.[7] This union occurred in a mythical time, when the first ruler of the town made an alliance with deities who gave him/her the power and control of the resources of the territory; in exchange for these favors, the ruler had to make offerings to them. As Fernández-Christlieb (2015, 341) has explained, the connection that communities felt toward their land was articulated in myth. This primordial bond made the ancestors inextricably tied to the *altepetl*. Sometimes the figures of the gods and the rulers merged to such a degree that it is impossible to separate them, as it is the case of *Ce Acatl Quetzalcoatl* (López-Austin 1973). On other occasions, the ruler's power was legitimized by listing the lineage that had preceded them, providing an account of how genealogies took root in the territory.

As can be seen, the *altepetl* reflected, through analogies and multiscalar replicas, the Indigenous concept of space (Figure 6.4). This was not merely a mappable geographic territory but was instead a complex integration of an *axis mundi* (either mountains or temples), mythical sites and elements where deities were present. It was not enough to paint the physical landscape to represent this complex spatial conception; in the images, rulers, ancestors, genealogies, toponymic

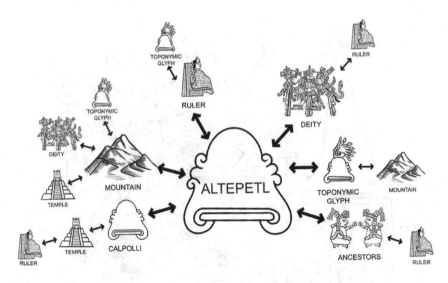

Figure 6.4 Diagram showing the different multiscalar relations of the *altepetl*.

Source: Drawing by Mariana Favila-Vázquez and Aban Flores-Morán.

glyphs, ancestral places, and temples had to be present, since they reflected and were the *altepetl*. The presence of these elements in the codices and later spatial documents reveal, just as deep maps do, the topological, relational, and narrative ties among the human agents and space. These ideas would persist in the paintings created after the arrival of the Spaniards. In the following section, we delve into the reconfiguration of this ideology and its representations in the Mesoamerican codices and the maps produced in the context of colonization.

Spatiotemporal representations in Mesoamerica and their transformations in the colonial period: mnemonic technologies shaping life and history

Writing "without words" as Hill Boone and Mignolo (1994) once put it, had a long tradition in Mesoamerica. Written since time immemorial, codices recorded vast quantities of information documenting cultural traditions, beliefs, religion, rites and ceremonies, genealogies, history, geography, mythical space, political alliances, time, and economic systems among many other topics related to life in Mesoamerica. This semantically rich pictographic technology evolved through expressions and codes directly linked to each society that created them, and although only a handful of prehispanic codices survived, aspects of this tradition continued up to the eighteenth century in colonial maps, *lienzos*, and codices (León-Portilla 2012).

The *tlahcuiloqueh*—in Nahuatl "the ones that write by painting"—could be men or women trained from an early age to record the histories and important information of the places that they rendered their service to (Galarza 1990). These could be temples, judicial courts, palaces, markets, or any other ceremonial or political center to which they belonged (Galarza 1980). As such, codices were not "authored" works for they belonged to the community and were thus not individually signed. The *tlahcuiloqueh* were educated in the *calmecac* (school) where they were trained, along with the nobility, in all cultural matters. The *tlacuilo* (sing.) recorded this vital information but may also have read these codices to the community and, on occasion, publicly displayed them. This meant that even common people could become familiar with many of the symbols and meanings contained within these records, which were kept locally in places called *amoxcalli* (library or the house of books) (Galarza 1992). Although it is known that these artists were in charge of creating these books, as pointed out by Torquemada, it might have been the priests that interpreted them (Torquemada 1969, 181).

As mnemonic technologies, the codices could serve multiple purposes and, as such, they might have recorded a plethora of topics. Unfortunately, only three types of prehispanic codices survived: calendric/ritual, historic/genealogic, and tributary. During Mesoamerican times, these documents served crucial social functions. The religious and calendric codices for instance, were employed to establish the dates for religious ceremonies or events of great importance such as the start of war, public works, among many others, as well as to indicate the festivities of the gods and times for rituals. As explained in the introduction, many of the codices contained spatial references, but these were understood in a substantially different way to European conceptions. While geographies are certainly expressed in these codices, in many instances, places could be intertwined with real and mythical time, history, and genealogies, as well as real and symbolic landscapes. In this sense, even with the classifications given by scholars, Mesoamerican codices can touch upon multiple "themes." Therefore, as Nowotny (1961) indicated in his now iconic work *"Tlacuilolli,"* to understand these themes, it is necessary to identify their multiple sections aiming to reveal how they relate to each other. The reading of the codices would usually be performed in the community, where oral history would be combined with these pictorial representations. This means that traditionally, using a combination of media, Mesoamerican cultures created in the codices powerful instruments of memory that immersed the reader and the listener into a deep space–time dimension, evoking directly or indirectly beliefs, customs, traditions, stories, historical accounts, and real and symbolic landscapes.

With the arrival of the Spanish, the systematic destruction of libraries and codices would begin, later continuing as *autos de fe* (a public penance), being deemed by the Catholic priests as works of the devil. Thousands of books and important libraries were destroyed. However, as early as the *Segunda Audiencia* (Second High Court) (1530–1535), codices were allowed to continue and were recognized as valid forms of recording. At the same time, some priests and chroniclers such as Andrés de Olmos, Bernardino de Sahagún, Toribio de Benavente (Motolinia), Diego Durán, Francisco de Burgoa, and Pedro Sánchez de Aguilar among others, would realize the value of these works for understanding native customs and systems. They aimed then to preserve or document, at least partially, the knowledge contained within these documents, using the surviving codices and the information provided by the *tlahcuiloqueh* to write their chronicles and histories.

From the second half of the sixteenth century, representations would increasingly take in European conceptions. For the codices from the colonial period, there are several "themes" that can be identified and classified as: historical; astronomical, calendrical and ritual; genealogic; cartographical and cadastral; economic, including censuses, financial and tributary records; ethnographic; natural history; religious; and a special type now called *Techialoyan* that served as official titles or papers linked to land tenure and ownership. This later type emerged toward the seventeenth century in Central Mexico due to the necessity of the native population to claim legitimacy and rights to their own lands and territories (León-Portilla 2012, 158).

After the establishment of the Spanish colonial power, knowledge depicted in the codices would continue, not only in the form of new colonial codices but also in the "transfer" of the contents of the surviving paintings into narrative forms of written history. This took place with the aid provided by the *tlahcuiloqueh* who were now in the service of the Church as artists, usually decorating the new temples, as well as by native informants and nobility. As León-Portilla has pointed out (2012, 145), this was the case of the *Relaciones* written by Chimalpáhin, the Florentine Codex by Fray Bernardino de Sahagún, and *Monarquía Indiana* by Fray Juan de Torquemada, among many others. In this manner, a great part of the documents produced during colonial times continued to be pictorial, but experienced strong transformations in terms of format, materials, painting techniques, and more specifically, in their purposes and themes.

While the Mesoamerican codices had portrayed the scientific knowledge, experience, narratives, and histories of the Indigenous kingdoms as well as their belief system, the themes of the new codices suffered radical transformations where new topics emerged, some disappeared, and others continued (Carrasco 1991). Some of the colonial codices are thought to be copies of Mesoamerican documents (e.g., *Tonalamatl de Aubin*, the *Matrícula de Tributos*, or the Codex *Borbonicus*), but the motives for their creation certainly changed, adapting to the necessities of the new reality, including the interest of the Crown and Church to know important details about the history, politics, economic system, and organization of the region. The creation of these documents was deemed necessary to better establish both the new colonial government and the Christian religion. Historical, economic, and judicial codices would become the most common themes, obeying the particular interest to record the history of a region (e.g., *Códice Azoyú, Códice Azcatitlan*), land ownership (e.g., *Códice de Santa María Asunción*), population censuses, all types of tribute and payments (e.g., *Matrícula de Huexotzingo, Códice de Otlazpan, Códice de Teloloapan*), and as support in judicial processes (e.g., *Códice Osuna*) (Glass and Robertson 1975; León-Portilla 1988).

The establishment of the viceroyalty and the processes that accompanied it, such as the granting of lands (*mercedes de tierra*), required the submission of detailed documentation which included requests for detailed paintings of the place(s) in question. In this way, a large and varied corpus of maps would amass over the course of the following centuries. In the same way as these colonial codices, the native cartographic tradition also continued to be drawn in the form of maps, but many retained Indigenous conceptions of space (see the second section). These representations included toponyms and other representations, but now could also integrate European elements including Spanish writing in the form of glosses, Christian symbols and ideas, new architectural elements such as churches, and European-style buildings, among others (e.g., the Map of *Tepecuacuilco*, the *Lienzo de Quetzpalan*, the maps I and II from the Kingsborough Codex, or those of the *Relaciones Geográficas*) (Bittman Simons 1974; Russo 2005; Mundy 1996). Aspects of the Mesoamerican tradition in colonial cartography would last until the eighteenth century. However, the most definitive changes observed in the sixteenth century can be summarized as the incremental disappearance of its pictographic form, giving way to the naturalism form of painting of the European Renaissance (Figure 6.5) (Escalante Gonzalbo 2010).

From our perspective, one of the aspects of great interest about the colonial era maps is that, although they were aligned to new interests and incorporated European beliefs, motives, styles, and conceptions in different ways, many of them would continue to function as native mnemonic technologies and continued to include traditional spatial and temporal representations, as well as ideological

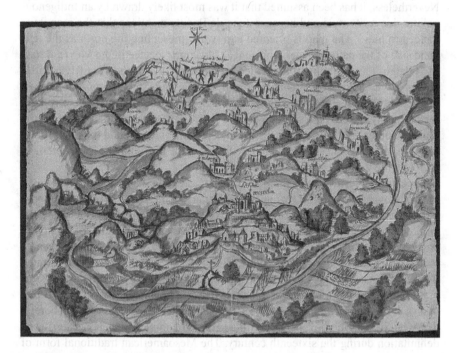

Figure 6.5 The Relación Geográfica de Meztitlán map.

Source: Courtesy of the Benson Latin American Collection, University of Texas at Austin (JGI xxiv-12).

renderings of the landscape that combined to form a profound representation of historical and environmental knowledge. In this manner, despite the variety of solutions used to represent the immense diversity of landscapes, many of these maps share much more than their simple administrative purpose or format (Russo 2005, 18). This is well exemplified in a great number of these maps, and here we will attempt to showcase this, using three colonial maps as examples of how these new representations emerged, sometimes preserving, sometimes combining, but in all cases offering expressions influenced by or considering past and present forms, carrying forward in a diversity of ways the depth and memory of the Mesoamerican tradition.

The reconfiguration of the Indigenous world: the maps of Atengo-Mixquiahuala (1579), Teozacoalco (1580), and Cempoala (1580)

Made around 1579, the map of Atengo-Mixquiahuala was presented by the *corregidor* (chief magistrer) Juan de Padilla to the viceroyal authorities as part of the questionnaire that would become the *Relaciones Geográficas*. The 1579 map is considered a "mixed" style due to the glyphs and representations in the Indigenous tradition and the inclusion of Spanish glosses and architectural depictions. Nevertheless, it has been assumed that it was most likely drawn by an Indigenous painter due to its style and because Juan de Padilla mentions that the map "was handed to him." The map is oriented east–west (presenting the north as the left-hand side of the painting) and represents a large area located in the Mexican state of Hidalgo. Part of the map is divided by the depiction of the Tula River and it covers the areas of Mixquiahuala, Tezontepeque, and Atengo, marked by their respective churches to the south of the river (Figure 6.6). Toponym glyphs in the native style were represented with the figure of the mountain, and they could allude to other *altepeme*. In this case, they are distributed around the map, showing, as with many other cartographic examples of the time, that the Mesoamerican pictographic writing tradition was still ongoing.

Located in a fertile valley, this Otomi area, as expressed by the *Relación*, was once a buoyant agricultural region and a tributary of the Triple Alliance. Its population was decimated as a result of the processes of colonization, including the enforced congregation of populaces, major disease outbreaks, and conversion to pastoral lands that provoked damaging environmental changes still noticeable today (Cook and Woodrow 1960; Pérez-Rocha and Tena 2000). As demonstrated by the work of Wright Carr (2009), the region underwent a profound transformation immediately after the conquest, where Mixquiahuala was subject to two main *encomiendas* (a labor system that rewarded the conquerors with laborers) that would be divided over time, with a series of complex changes in territorial delimitation during the sixteenth century. The Mesoamerican traditional form of government, presided by lords that would rule over their lifetime, eventually gave way during the colonial period to elected governors that led the native councils. Life and events at this time, but particularly those related to issues of justice and

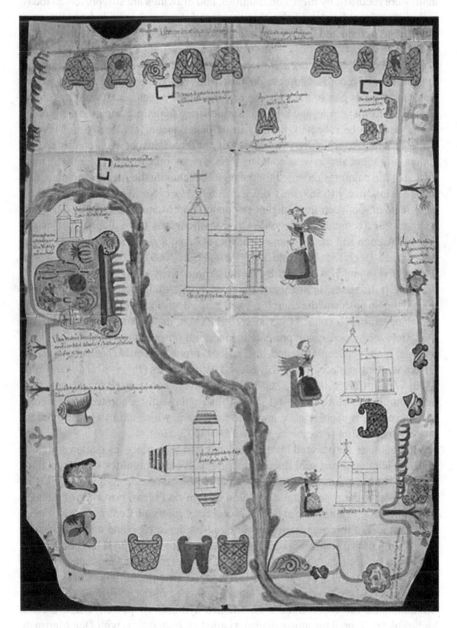

Figure 6.6 The Relación Geográfica de Atengo-Mixquiahuala map.

Source: Courtesy of the Benson Latin American Collection, University of Texas at Austin (JGI xxiii-12).

land, were recorded by these communities, and drawings are still preserved today, including primordial land titles that are used by a number of Indigenous communities to defend and make legal claims about their lands (Ruiz Medrano 2001, 211–18).

Snippets of events where *cabildos* (councils) were involved can be also explored in the historical records. Closer to the time when the Atengo-Mixquiahuala *pintura* was submitted as part of the RGs, the region records a recurrent form of resistance to Spanish forms of rule, and the abuses which they sometimes incurred. This opposition seems to be reflected not only in the legal documents located in Mexico's National Archive, which record a series of conflicts between the Spanish and the communities, but also in what could be seen as mnemonic forms of resistance, manifested in the paintings recorded for the region, including the Atengo-Mixquiahuala map. For instance, in 1569, the *Códice de Santa María Nativitas de Atengo* records an agreement between the Indigenous council and Manuel de Olvera, son of the *conquistador* (conqueror) Diego de Olvera, whom was *Corregidor* (chief magistrate) of Atengo and *alcalde mayor* (mayor) of Mixquiahuala at the time (Dorantes de Carranza 1999, 155; Sterpone 2001; Wright Carr 2009). The document was written by Olvera and was painted by an Indigenous artist in the native pictographic tradition. It discusses, in Spanish text, the economic obligations of the community toward the Spanish and the church, as well as the distribution of the tribute gathered and managed by the council (Sterpone 2001). On it, the native governor of Atengo, *Don Martin de Porras*, is represented along with other Indigenous officials. Only one year later, between 1570 and 1572, a legal process recorded against Manuel de Olvera was put in place by possibly several native nobles (*Proceso de Oficio* 1570–1572; Wright Carr 2009, 27–32). Manuel de Olvera had been taking advantage of the communities since at least 1569 and had unlawfully requested and received several payments in coin, labor, and food from them, which ultimately triggered the process against him (*Proceso de Oficio* 1570–1572, 59v).

While abuses perpetrated by the Spanish often took place, as demonstrated by both the legal and pictorial documents, the native communities rarely remained passive. A really interesting aspect of the *Códice de Santa María Nativitas de Atengo* is that the representation of Don Martin de Porras, governor of Atengo, was done in the traditional style for Indigenous rulers of Central Mexico: he is seated on a reed throne, wearing a white red-bordered cape and a loincloth, with no sign of Spanish status markers displayed. As Olko (2014, 243) has pointed out, this could be seen as "rather anachronistic considering that the manuscript was painted around 1569." In a similar way, the three prominent figures of the native rulers from Mixquiahuala, Atengo, and Tezontepec in the 1579 map of Atengo-Mixquiahuala are represented with Mexica symbols of power. These include high-rank traditional turquoise diadems called *xiuhuitzolli*; as with Don Martin de Porras in the *Santa María* codex, the rulers are sitting in *tepotzoicpalli*; and they are marked as exceptional warriors by their headdress with a prominent ribbon called *quauhyacatl*. In addition, an interesting aspect of the ruler of Atengo is that the style of his dress is one of the few examples outside of the Valley of Mexico or conquered areas by the Mexica of a royal *manta* (cape) decorated with a turquoise mosaic—the *xiuhtlalpilli tilmatli tenixyo* (Olko 2006, 144). This *manta* is considered as the most prestigious royal cape used by the Mexica rulers, and although it has been explained by Anawalt (1997) that this region did produce this type of *manta*, the presence of other prestigious Mexica symbols seem to indicate the recall of an imperial convention more than a local tradition (Olko 2006, 145). In any case, what is of great interest here is the Mesoamerican cultural and political connections that are portrayed by these representations. The Otomi rulers of these lands spoke Nahuatl as a second language (Wright Carr 2009, 30), and as a tributary of the Mexica adopted their iconographic conventions which certainly continued to be seen as forms of power during a great part of the colonial period.

While the presence of Indigenous rulers in the colonial maps could be expressions of former rulership sometimes related to the founders of these towns (Rodríguez Cano and Torres Rodríguez 2001; Acuña 1985), in the case of this region, we propose that a form of mnemonic resistance was taking place through these representations. These documents would, in effect, have served a Spanish intent, but by remembering and highlighting the nobility's connection to traditional forms of prestige and their deep symbolism, these documents communicated, in the native way, the power that rulers previously had and still possessed. It must be remembered that Mesoamerican rulers were held in a sacred light (Ruiz Medrano 2001, 120–1), and as previously explained, their connection to land had a mythical origin. While the Spaniard Manuel de Olvera might have written the agreement presented in the *Códice de Santa María Nativitas de Atengo*, the native governor *Don Martín de Porras* was represented with the symbolism of the Mesoamerican rulers as an equal. In effect, although in the case of the map of Atengo-Mixquiahuala it is not clear whether the Otomi rulers were historical or current ones, the symbolism used invokes the power of the Mesoamerican ruler, bringing to mind this kind of resistance and continuation.

During the colonial period, recalling the past and history would become a form of defending lands. Possibly related to this, in addition to the introduction of European concepts, was a fundamental shift in the representation of space. As explained in detail elsewhere (Murrieta-Flores et al. 2019), the map of Atengo-Mixquiahuala shows its locations with remarkable precision. It has been argued that the map might have existed before, or that there might have been a copy already in the community, given that Juan de Padilla says that it was granted to him. Nevertheless, an interesting aspect of the description given in the *Relación* by the *corregidor* is that he fails to answer some of the questions posed by the royal questionnaire, but requests that the reader directly refers to the *pintura* (e.g., the response to questions 4, 10, and 18). Question 18, for instance, asks for details of the positions of each town and the surrounding topography, and he does not describe this, asking the reader to instead check the painting, even saying that this information can be seen better there (Acuña 1985). This means that, although he might not have expressly shown in the text knowledge of the toponyms or other symbolic meanings present on the painting, he certainly recognized the geographies depicted as accurate, to the extent that he did not deem it necessary to add further explanation to these questions. In a sense, Padilla deemed the *pintura* as useful for this purpose which, accompanied by his textual description, was adequate to fulfil the royal mandate. This might indicate that toward the end of the sixteenth century, a shift in the conception of space had already taken place, giving way to a "Cartesian" idea that the Spanish could not only easily recognize but that might even have been thought as better than a textual description.

The map, nonetheless, would still maintain important elements of the picto-graphic native tradition, including the strong symbolic weight of a religious land-scape. The conversion to Christianity and the condemnation of existing religion would have a profound impact and a crucial role. Yet, at least toward the end of the sixteenth century, elements of a combined symbolism can still be found in the *pinturas*. Christian churches would assume focus in Indigenous spatial rep-resentations as the new significant landscape markers and these would become the spatial references *par excellence*. Russo (2005, 48–9) has argued that while the Spanish representations of towns in their maps would usually allude to the governmental palace more often than to a Christian building (similar to some of the heraldic conventions), the Indigenous paintings took the cross, using it as a locational reference even more than as a religious one. While there is no doubt regarding the use of churches as new landmarks, we think that there is also a continuation of the association of symbolic spaces and traditional sacred land-scapes in many of these paintings. The map of Atengo-Mixquiahuala, for instance, clearly portrays the churches as significant features. However, their association with the native rulers might speak also of the connection between them and the new religious temple, where the ancient deities were now embodied in the figure of the Christian saints, still maintaining a close relation to the concept of the *altepetl*, as explained in the previous section. Furthermore, although not in its immediate or jurisdictional territory, the Atengo-Mixquiahuala map also recalls the presence of Tula and Cerro Xicuco (Figure 6.7). This is not of small signifi-cance given that Tula is the ancient city of Tollan Xicocotitlan ("the Tollan next to Xicuco Hill") that held enormous mythical and religious significance across Mesoamerica as the place associated with the god Quetzalcoatl. This association, however, might have taken a double symbolic undertone in the sixteenth century as, with the arrival of the Spanish, the figure of Quetzalcoatl would acquire (for a series of complex reasons) a Christianised connotation (López-Austin 1998).

The Christian influence would be evident not only in the portrayal of churches and monasteries but also in the spatial arrangement of some of the maps and motives, where the possible influence of the European conception of time can be seen. An interesting example is the Mixtec map of Teozacoalco (Figure 6.8). Dating from around 1580 and related to the Mixtec community of San Pedro Teozacoalco in the modern state of Oaxaca, this map was also drawn by an Indig-enous artist as part of the response to the RGs questionnaire. The map combines European conventions with Mixtec codex–style representations. The map shows, in a circular depiction, the extent of the colonial territory of Teozacoalco including boundary markers, natural features, pathways, towns, churches, and toponyms, as well as Spanish glosses, and on the left-hand side, the representation of two genealogies in the form of columns (Anders et al. 1992, 24–6) (Figure 6.9). First

San.tamaria deatongo

Figure 6.7 Tula is represented on the lower-right end of the map (identified with the gloss) while the toponym of Cerro Xicuco is shown on the right-hand side of the church.

studied in depth by Alfonso Caso, this map and its genealogies would serve, as he expressed, as the "Rosetta Stone" that would allow him to establish a proposed chronology for Mixtec history (Caso 1949).[8] Containing genealogic representations of individual pairs of founders connected with Spanish glosses in the RG map, Caso identified these as the rulers of Teozacoalco and Tilantongo, making the link between these and the representation of the prehispanic dynasties in the codices Zouche-Nuttall and Vindobonensis (Nuttall, 1902). This allowed him to propose a correlation of the Mixtec historical chronology with the Christian calendar, unleashing a myriad of interpretations from the scenes in the codices,

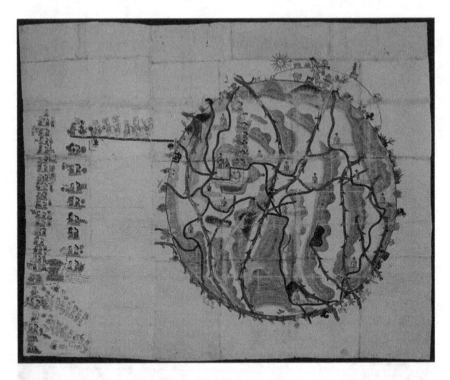

Figure 6.8 The Relación Geográfica of Teozacoalco map.

Source: Courtesy of the Benson Latin American Collection, University of Texas at Austin (JGI xxv-03).

and opening important venues to advance the understanding of the Mesoamerican calendar system (Jansen and Pérez Jiménez 2011, 112).

It is important to point out that the presence of these kinds of genealogies are not common in the maps of the RGs and that, while the representations of the dynasties in the Teozacoalco map use the iconographic Indigenous tradition, it could be argued that the way these are displayed may in fact be obeying the European conception of time. The questions asked by the RGs such as who discovered and conquered their land (Q.2), who were the rulers before this (Q.14), how they were governed (Q.15), and what were the good or evil practices they sustained (Q.14), made the Indigenous informants consider and portray their own history in strict relation to the Conquest, the Christian world, but most importantly in a linear way, which was completely different to how Mesoamerican communities conceived history and time. Contrary to the linear nature of European time, the native conception of it was cyclical. This had important implications in terms not only of cosmogony but also for the expressions of history and its narratives, as can be observed in the Mesoamerican codices. Although Indigenous societies indeed recorded and recognized individual historical events in order to mark acts of

Figure 6.9 Detail of the starting point of the genealogies of the rulers of Teozacoalco and Tilantongo (lower-left corner of the map). The genealogy starts with a toponymic glyph with the Sky Temple represented with a blue roof with star eyelets on top a base with black frets (a). The married couple facing each other are *Humo que Baja del Cielo* (Smoke that descends from heaven) (b) and *Quetzal Precioso* (Precious Quetzal) (c). Their names are given in the form of hieroglyphs next to them.

origin, conquest, or rights among others, these were subordinate to the cyclical nature of time, where these events could take on a new light intertwining with new knowledge, but also enter the mythical realm in a variety of ways (Gruzinski 1991, 83). As we see in the calculations made by some of the elders and informants in the *Relaciones Geográficas*, the native informants made the conversions of time asked by the Spanish, complying with the request to provide dates for historical events. Nevertheless, although both systems could be "translated," this did not change the fact that they expressed radically different views. The Spanish placed an emphasis on "pinning down" dates, chronological recording, and measurement, while Mesoamericans focused on the nature and quality of the moment. As Gruzinski (1991, 84) brilliantly put it, the native date of "Ce-Tecpatl, 1 Silex" for example, "did not refer to a past century or to a singular time span, but to a type of year, to a range of received influences, or to other years of similar name." In this manner, dynastic histories would divide time in a particular way, acquiring its own characteristics and essence. These histories were told, often surrounded by religious and mythical events, and could be presented as independent units and not a successive narrative, as is the case with the Zouche-Nuttall codex. This is beautifully expressed in folio 37 (and 38) of this codex, where the emergence of Apoala is narrated (Figure 6.10).

The two foundational couples are Lord 1 Flower and Lady 13 Flower sitting on the left (Figure 6.10), who are the parents of Lady 9 Alligator who sits with her husband, Lord 5 Wind-Rain on the right. They sit on top of the two rivers in the region, and a procession of four priests takes place above. These go in peregrination carrying multiple sacrificial items and enter the cave known as Yahui coo maa (represented with the toponym of the open mouth feathered serpent), and into the realm of the god of rain to make offerings (portrayed in the next folio) (Anders et al. 1992, 165–8). The foundational story takes place in a location that has real geographic correspondence with that of the Apoala Valley where the rivers, the cave, and the waterfall have been identified alongside other key elements (Anders et al. 1992, 39–53). In the codex, however, these are also intertwined with history, toponymy, and mythical space and time. In contrast, the couples of the genealogies of the Teozacoalco map are not embedded in this richer context and are presented in a linear fashion, adapting, in our view, to a format that could make better sense to the Spanish, adding a brief explanation in handwriting and giving the names of the surviving heirs of the dynasties at the time of the creation of the map. A similar arrangement is shown in the *Lienzo de Nativitas* from Coixtlahuaca, also in Oaxaca.

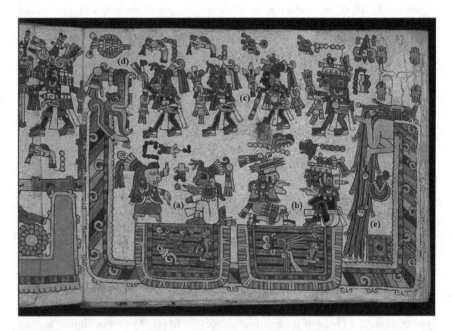

Figure 6.10 Representation of the rulers of the Apoala Valley and its geographic and sacred landscape in the Zouche-Nuttall Codex, folio 37.

Source: Image courtesy of the British Museum.

Another important characteristic of this painting is its circular representation. While some scholars such as Smith (1973) and König (2011) have argued that this comes from the European tradition responding to an Indigenous adaptation of the medieval mappa mundi, Mundy (1996) has proposed that it constitutes a "communicentric" projection, placing the community as the center. In this case, the circular convention may have reflected a conception of the landscape, where the surroundings were arranged in a circle around a central town, as it is conceived by the Pueblo Indians (Mundy 1996, 116–17). An aspect of great interest here is the interplay between time, power, and the detailed depiction of the territorial boundaries. On the one hand, the remote past is presented in the genealogies outside the main circle that are connected to the toponyms of Teozacoalco and Tilantongo, which interestingly, are combined with the names of their prehispanic ceremonial centers (Anders et al. 1992, 26). This seems to be recalling the mythical-foundational act, declaring the sacred nature of the land, where the connection or alliance with the divine enables the appropriation of the territory (Carmagnani 1988, 15). Furthermore, in the group of figures connecting the genealogic section with the circular part, the depiction of Lord 2 Dog throwing an arrow toward Teozacoalco, represents the symbolic act of the *toma de posesión* (taking possession of the land) (Oudijk 2002, 108) (Figure 6.11). On the other hand, generations closer to the time when the map is created are depicted next to the colonial church of Teozacoalco and inside the circle that shows the territorial boundaries.

Although the map might have been created in response to the Spanish questionnaire and with a European audience in mind, this interplay between the delimitation of the land, the presence of the foundational couples, and the action of rulers such as Lord 2 Dog in connection to the sacred past and present, seem to continue to express the core elements of the *altepetl*, where components such

Figure 6.11 Lord 2 Dog is represented as a Chichimeca warrior with bow and arrows in his hand while confronting the people of Teozacoalco. Symbolic or real wars were legitimate forms of taking possession of the land. In this scene, it is plausible we are witnessing precisely this act (Oudijk 2002).

as the *toma de posesión*, can be traced back to the eleventh century and have maintained an astonishing continuation up to the twentieth century in Oaxaca (Oudijk 2002, 108–9). Furthermore, we see these historical figures walking into Teozacoalco and the rest of the territory. The act of ritual walking around the land took place in order to take possession of it, connecting land to people, and this practice of the *toma de posesión* still takes place today (García Zambrano 2006). By recalling the foundational act and history and with its detailed definition of territorial boundaries, it reminds us of the use of other later *lienzos* in important symbolic and territorial practices, possibly similar to what is explained for the *Lienzo de Petlacala* in the conclusion to this chapter (see Jiménez Padilla and Villela Flores 2003).

A final example that warrants attention as to the depth of the symbolic representations related to the prehispanic conception of space in colonial maps is the case of the painting accompanying the *Relación* of Cempoala (1580). This painting represents a large geographic area filled with Indigenous toponym glyphs, religious buildings (mostly parish churches), rulers, and landscape representations (Figure 6.12).

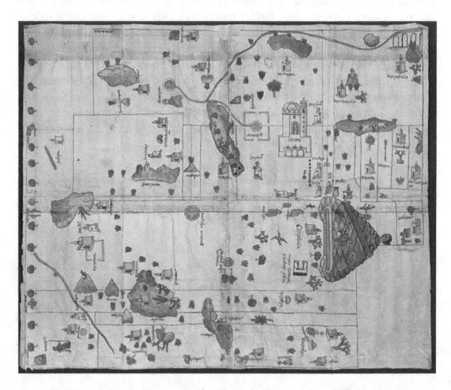

Figure 6.12 The Relación Geográfica of Cempoala map.

Source: Courtesy of the Benson Latin American Collection, University of Texas at Austin (JGI xxv-10).

Figure 6.12 presents the territories affiliated to the town of Cempoala (modern Zempoala, Hidalgo) and its neighbor Epazoyucan (Hidalgo). The red lines that create rectangles all over the surface have been interpreted as the divisions of the territories corresponding to each main town/*altepetl* (Mundy 1996, 129). Possibly one of the most prominent features in the map is the largest toponym glyph, which, since this document accompanied the Relación of Cempoala, it is not surprising that it corresponds precisely to this town (Figure 6.13). Nevertheless, what is of great interest is that this toponym has characteristics that recall the prehispanic elements of the altepetl in that the figure of the mountain with a tucked base and its surface depicts the rhomboidal pattern associated with the earth monster and at its summit appears a human head interpreted as a Totonac, with whom the Mexica associated Cempoala (Mundy 1996, 142).

Although the representation does not really have any feature that allows us to identify the ethnic affiliation of this figure,[9] ornaments such as a circular "bezote" under the mouth, circular pendants known as "orejeras," as well as a red ribbon in his hair present evidence of a high-ranking personage usually associated with warriors, rulers, and merchants. The components of the glyph, however, do not allow the reading of the toponym, since Cempoala means "place of twenty" which

Figure 6.13 The toponym of Cempoala.

refers to the establishment of a market every 20 days (*Cempohualli* in nahuatl means "twenty"). The presence of the human head as a part of the glyph, there-fore, could be explained as an integration or connection of important groups of the community to the spatial concept of *altepetl*. Interestingly, Escalante Gonzalbo (2010, 361) has pointed out that, despite the extensive degree of experimenta-tion that native painters carried out during the sixteenth century, glyphs proved especially resistant to change. While native artists experimented with the human figure and architecture in general, it has been argued that they rarely abandoned the pictographic representation of particular objects such as water and mountains (Escalante Gonzalbo and Flores Morán 2020, 77). In a way, the long continuation of the representation of toponymic glyphs, and particularly that of the mountain is not surprising, given its pivotal role in Mesoamerican cosmogony and the spatial ontological configuration of the *altepetl*.

Conclusion

In 1577, when the royal chronicler–cosmographer Juan López de Velasco requested maps to be sent accompanying the answers to the questionnaire that each town should provide across the viceroyalty, and that would become the basis for the now invaluable collection of the Geographic Reports of New Spain, he most likely did not expect to receive geographic information in the form of the *pinturas* he was sent. The questions asked by the Spanish were probably similarly received by the Indigenous elders, because the RGs questionnaire asked the informants to think in a frame of reference and of categories that were unlike those with which these elders most likely conceived of the world (take for instance the magic/religious conceptions about the territory and the natural world as explained in the second section) (Gruzinski 1991, 80). Although the process of Christianization had been ongoing for some 40 years at this point, this happened at different rates in diverse regions. According to sources around the northeast of Oaxaca, for instance, the elders interviewed were between 75 and 90 years old. This means that they had been born and raised in the Mesoamerican tradition, and by the time of the RGs, were possibly the last generation remnant before contact with the Western world (Gruzinski 1991, 81). Despite the massive destruction of the codices and the intro-duction of a different way of writing, the oral tradition and inherent knowledge that went hand in hand with the *pinturas* was still in memory and, in many cases, alive. This is shown across the documents of the *Relaciones Geográficas* in a number of ways: sometimes explicitly acknowledging and lamenting the loss both of the *pinturas* and the elders that knew the tradition (e.g., RGs of Atlatlauca y Malin-altepeque, Oaxaca; Tlaxcala; Texcoco); on other occasions, using the mention of the *pinturas* and the recounting of them by the elders as a form of validation of the information recorded in the *Relacion* (e.g., RGs of Itztetepexic; Tilantongo, Oaxaca; Meztitlan, México; Querétaro, Michoacán); and others explaining the many uses the *pinturas* could have, including as historical records (e.g., RGs of Acatlan, Tlaxcala; Atlatlauhca; Coatepec; Chimalhuacan Atoyac; Chicoaloapan; and Meztitlan, Mex; Siróndaro, Mich.), to forecast the weather, and for calendric or divination purposes (e.g., RG of Atitlán, San Andrés, Guatemala; Epazoyuca, Mex.), among many others. According to what the RGs record, therefore, by the end of the sixteenth century it is clear that while in some regions the Mesoamerican tradition seems to have faded, in many others it had not.

It is certain that many of the Mesoamerican spatial conceptions might have been lost or suffered transformations during the early colonial period with the introduction of the new religion, social reconfiguration, and the loss of oral and interpretative knowledge due to the devastation of disease outbreaks. In the spatial representations of the *Relaciones*, however, we can observe forms of adaptation, recasting and adoption of new conceptions, but also of persistence, as well as a historical and symbolic recalling of Mesoamerican formulations. These Indigenous mnemonic technologies of the "New World" would continue to exist, and it can be argued that, despite the transformations, they continued expressing spatiotemporal, historical, and religious dimensions of enormous significance that Indigenous communities relentlessly engaged with. The use of these technologies persists even today, taking special importance and significance in Mexico's modern political climate. Native communities across the country have used their primordial titles and *pinturas* to defend their lands (not always successfully), from the sometimes-systematic attempts by industry and the government to dispossess them (Ruiz Medrano 2001, 211–74). Codices and lienzos with spatial information still possess historical, religious, and symbolic significance in many of the communities. One of many examples is the *Lienzo de Petlacala* (Jiménez Padilla and Villela Flores 1999, 58–61, 2003; Villela Flores 2018). This *lienzo* is a document dated possibly from the seventeenth century located in the Nahua Mountain region of Guerrero. It is composed of a central panel with borders that depicts the history of a Nahua migration from the Valley of Mexico that was carried out in the middle of the fifteenth century, culminating in the foundation of the actual town recorded by toponyms and glyphs, as well as the demarcation of its limits. In the central panel, the figure of Emperor Charles V is displayed in front of three native nobles, while below a kneeling woman is portrayed inside the church. The locals identify this woman as the founder María Nicolasa Jacinta, and the inhabitants of Petlacala relate how she and Charles V, along with the nobles, walked together throughout the region to establish the boundaries of the town (Jiménez Padilla and Villela Flores 1999, 60). Charles V and Nicolasa Jacinta are thought to be the founders of this town, and this mythical relationship seems to have granted a sacred character to the *lienzo*, which the community regards as an item of power and it is used in their yearly ritual cycle (Figure 6.14).

The *lienzo* precedes, for instance, the municipal election and particularly the act of the *toma de posesión*, legitimizing the transfer of rulership to the new representative through the communal power that the original founders in the *lienzo* exert. This seems to be linked to a possible continuation of the Mesoamerican conception of power where, as previously noted, its legitimation was intimately associated with genealogical relationships, land, and ancestral legacy. The combination of the presence of the mythical and real founders with the spatial aspect of the *lienzo* creates a hierophany. The demarcation of territorial limits and the

Figure 6.14 A member of the community presents an offering to the *Lienzo de Petlacala*. This is displayed across the year in the family altar of the municipal commissioner (Jiménez Padilla and Villela Flores 1999).

Source: Photo courtesy of Samuel Villela Flores.

act of walking the boundaries creates a rupture in the continuity of profane space and thereby sacralizing it (Jiménez Padilla and Villela Flores 2003, 104–5). The narration of the founders walking through the territorial boundaries constitute the recollection of the primordial foundational act, that had even more centrality than the construction of the church (Gruzinski 1991, 124).

Furthermore, once a year, the *lienzo* is taken to an altar in the top of *Petlacalte-petl* hill, just before the start of the rainy season, where a ceremony takes place and a "petition for water" is made (Figure 6.15). The researchers that have studied these rituals, Jiménez Padilla and Villela Flores (1999, 2003), have identified the existence of at least 60 pictorial documents of the so-called historical–cartographic tradition in the communities of this region. A matter of great interest here is that the possible continuation of the core concept of the *altepetl* can be observed in these practices, where an association between the mountain, the mythical foundation story and its rulers, as well as the sacred and natural world, takes a living dimension through the painting, in combination with Christianized practices.

Throughout this chapter, we have sought to showcase how spatial technologies and conceptions in the Indigenous world were conceived, while explaining some of the factors for their transformations during the colonial period. In doing so, we also hope to make clear our position by voicing that these expressions

Figure 6.15 The *lienzo* is placed in the altar at the top of Chichitépetl Hill for the petition
of rain. The depiction of the hill is included in the spatial section of the *lienzo*,
and given the archaeological remains, this seems to be the original prehispanic
site of occupation (S. Villela Flores, personal communication).

Source: Photo courtesy of Samuel Villela Flores.

should not be labeled as purely cartographic products, as they often have been.
This is particularly important considering the uptake of tools such as Geographic
Information Systems (GIS) within Latin American archaeology during the last
10 years. While the application of advanced spatial analysis with GIS can aid our
understanding of certain phenomena, historical archaeologists need to be aware
of the conceptual lens that these tools impose. In the field of Spatial Humanities,
deep mapping emerged as a form of addressing the shortcomings that the use of
this kind of technology can have. Nevertheless, it must be considered that tools
like GIS were never created to analyze the complex systems that human cogni-
tion, emotion, and belief constitute. In a world where the rise of computational
approaches, data mining, and pattern identification are increasingly valued, it is
interesting to see from a decolonial Latin American perspective how, much like
Mesoamerican time, the Spatial Humanities have come full circle. After walking
hand in hand with GIS, the field is coming back to, or perhaps more accurately,
increasingly proclaiming the idea that spaces are fluid: they connect people but
also define them and are social, cultural and historical palimpsests interlacing
the collective experience of a multitude of generations, their times, and voices. If
our aim is to understand these spaces, therefore, they can hardly be confined to
Euclidean cartographic representations.

The meaning and value of Deep Mapping as a theoretical and methodological approach has been discussed in a multiplicity of works (including the series from which this final book emerges), at times being praised, and other times criticized due to its varied methods and practitioners (Ridge et al. 2013; Bodenhamer et al. 2015; Harris 2016; Roberts 2016). What the current of this field brings, however, is that while embracing the benefits of technology, it has also provided the ground to discuss what "space" and "place" mean in the Digital Humanities, and how, despite technological constraints, there is certainly value in the exploration of digital methods. The crucial problem from our perspective, however, is that the development of these approaches—just like GIS—usually takes place in the context of Western paradigms. In the case of Mesoamerican and Colonial History and Archaeology, the application of advanced computational methods for their study is still in its infancy, and this can be seen as an advantage. In the case of Latin America, and possibly the majority of the so-called Global South, the use of digital methods, tools and technologies for the study of the past need to be specially approached with a critical lens, given the divergent and even opposite worldviews that some cultures can have to that of the West. In fact, there is a strong case to be made to imagine and develop decolonial tools, technologies, and approaches. Through the case presented in this chapter, we hope to have made the reasons for this explicit. While 500 years have passed since the encounter between Indigenous societies in this region and Europeans, the cultures emerging from the aftermath of the encounter did not simply forget their ancient Mesoamerican heritage and traditions. With this, we do not mean that these societies have remained static or fixated in the past, but that the processes, traditions, customs, and beliefs that inform identity can endure despite their transformations. Beyond the weighty modern connotation that the term sometimes entails, the historical documents analyzed here can certainly be thought of in the wider sense as maps. Conceptually, they portray spatial relationships and their geographies can sometimes be analyzed from a cartographic perspective. Our aim here is not to deny the role of the Western tradition in the production of these paintings, but to point out that in its syncretic process, they go far beyond "simple" cartographic maps. These paintings acted—and in instances such as that of Petlacala, continue to act—as dynamic repositories of time and knowledge, instruments of history, legitimation and sacred devices, and holders of power and continuation for these communities.

Acknowledgments

We would like to give special thanks to Samuel Villela Flores and Pablo Escalante Gonzalbo who have generously shared their knowledge and research with us, and to Katherine Bellamy for her thoughts and corrections to the manuscript. This research was supported by a grant from the Transatlantic Platform of Social Sciences and the Humanities (Digging into Data) through ESRC-UK, CONACyT-Mexico, and FCE-Portugal. Grant number ES/R003890/1: "Digging into Early Colonial Mexico: A large scale computational analysis of 16th century historical sources" (DECM).

Notes

1 In the RG questionnaire, López de Velasco asked for precise distances and specific directions to and from towns and *cabeceras*. The royal chronicler–cosmographer might have expected to receive paintings with the cartographic techniques that have been developed to administer lands, and that, as Fernández-Christlieb (2015, 336) explains, had been fostered by the "activity of *Landschaft* painters within the Iberian Peninsula" sponsored by the Spanish crown.

2 The organization of these overlapping levels with a static nature has recently been discussed from a new approach to primary sources that suggests a more dynamic characterization in which human and nonhuman agents influence the essence of these places (see Díaz et al. 2015).

3 The myth explains the origin of the earth by describing how the gods Quetzalcoatl and Tezcatlipoca separated Cipactli's body in two parts: one part would constitute the surface and the other the sky (López-Austin 1997, 92–3).

4 The equivalent of this concept in other Mesoamerican areas are the *tayu* in the Mixtec tradition (Hirth 2012, 74), and according to Restall (1997) the colonial Maya term *cah*.

5 The word *altepetl* is a portmanteau; this is a single metaphorical concept created from two different words. In this case, its origin is the conjunction of the terms in Nahuatl *atl*, "water," and *tepetl*, "hill or mountain." The translation would be "water-mountain" or "mountain of water" (Contel 2016, 89).

6 According to Escalante (2010, 19) the Nahuatl system of writing is pictographic, that is to say, "It is a language of painted or carved forms that follows certain stylistic conventions. But it is a *sui generis* pictorial language, since it seeks not only the visual representation of objects, characters, situations and ideas, but also the precise recording of events and data."

7 This can be seen in the Stone of Tizoc, a basalt monument from the fifteenth century depicting Mexica warriors taking captive the rulers from different places.

8 Caso worked as an archaeologist in the iconic site of *Monte Albán* and at the time had increasingly studied the iconography of several Mesoamerican codices.

9 The identification of the individual as a Totonac comes from the comparison to the toponym of Cempoala depicted in the Codex Mendoza (Mundy 1996, 129). This interpretation was originally made by Antonio Peñafiel (1885).

References

Acuña, René. 1985. *Relaciones Geográficas del Siglo XVI: México*. Vol. 6, 10 vols. México: UNAM.

Anawalt, Patricia Rieff. 1997. 'A Comparative Analysis of the Costumes and Accounterments of the Codex Mendoza'. In *The Essential Codex Mendoza*, edited by Frances Berdan and Patricia Rieff Anawalt. Berkeley: University of California Press.

Anders, Ferdinand, Maarten Jansen, and Gabina Pérez Jiménez. 1992. *Crónica Mixteca. El rey 8 Venado, Garra de Jaguar, y la dinastía de Teozacualco-Zaachila. Códice Zouche-Nuttall: ms. 39671 British Museum, Londres*. Códices mexicanos. México: Fondo de Cultura Económica.

Bittmann Simons, Bente. 1974. 'Further Notes on the Map of Tepecoacuilco, a Pictorial Manuscript from the State of Guerrero, Mexico'. *Indiana* (2): 97–132.

Bodenhamer, David J., John Corrigan, and Trevor M. Harris. 2015. *Deep Maps and Spatial Narratives*. Bloomington, IN: Indiana University Press.

Broda, Johanna. 1991. 'Cosmovisión y Observación de La Naturaleza. El Ejemplo Del Culto de Los Cerros En Mesoamérica'. In *Astronomía y Arqueoastronomía En Mesoamérica*, edited by Johanna Broda, Stanislaw Iwaniszewski, and Laura Maupomé,

461–500. México: Instituto de Investigaciones Históricas, Universidad Nacional Autónoma de México.

Broda, Johanna. 2015. 'Political Expansion and the Creation of Ritual Landscapes: A Comparative Study of Inca and Aztec Cosmovision'. *Cambridge Archaeological Journal* 25: 219–38.

Carmagnani, Marcello. 1988. *El Regreso de Los Dioses: El Proceso de Reconstitución de La Identidad Étnica En Oaxaca, Siglos XVII y XVIII.* 1. ed. Sección de Obras de Historia. México: Fondo de Cultura Económica.

Carrasco, Pedro. 1991. 'La Transformación de La Cultura Indígena Durante La Colonia'. In *Los Pueblos de Indios y Las Comunidades.* 1st ed., Vol. 2, edited by Alicia Hernández Chávez and Manuel Miño Grijalva, 1–29. Lecturas de Historia Mexicana. México: El Colegio de México.

Caso, Alfonso. 1949. 'El Mapa de Teozacualco'. *Cuadernos Americanos* 8 (5): 141–81.

Castañeda de la Paz, María and Michel Oudijk. 2011. 'La Cartografía de Tradición Indígena'. In *Historia General Ilustrada Del Estado de México.* Vol. 2, 87–112. Zinacantepec, México: El Colegio Mexiquense, A.C.; Gobierno del Estado de México; Poder Judicial del Estado de México; LVII Legislatura del Estado de México.

Cline, Howard. 1972. 'A Census of the Relaciones Geográficas of New Spain, 1579–1612'. In *Guide to Ethnohistorical Sources, Part One.* Vol. 12, 324–69. Handbook of Middle American Indians. Austin, TX: University of Texas Press.

'Codex Borbonicus'. n.d. Accessed 14 July 2020. www.famsi.org/research/graz/borbonicus/thumbs_1.html.

'Codex Borgia'. n.d. Accessed 14 July 2020. https://digi.vatlib.it/view/MSS_Borg.mess.1.

'Codex Boturini (Also Known as Tira de La Peregrinación)'. First Half of 16th Century. Accessed 14 July 2020. https://commons.wikimedia.org/wiki/File:Boturini_Codex_(folio_5).JPG.

Contel, José. 2016. 'Tlalloc-Tlallocan: El Altepetl Arquetípico'. *Americae* 1: 89–103.

Cook, Sherburne Friend and Woodrow Borah. 1960. *The Indian Population of Central Mexico, 1531–1610.* Berkeley: University of California Press.

Cortés, Hernán. 1856. *Cartas y Relaciones de Hernán Cortés al Emperador Carlos V.* Edited by Pascual de Gayangos. París: Imprenta Central de los Ferro-Carriles A. Chaix y ca.

Díaz Álvarez, Ana María, coord. 2015. *Cielos e inframundos: una revisión de las cosmologías mesoamericanas.* México: Universidad Nacional Autónoma de México, Instituto de Investigaciones Históricas.

Dorantes de Carranza, Baltasar. 1999. *Sumaria relación de las cosas de la nueva España: con noticia individual de los conquistadores y primeros pobladores españoles.* México, DF: Editorial Porrúa.

Escalante Gonzalbo, Pablo. 2010. *Los códices mesoamericanos antes y después de la conquista española: historia de un lenguaje pictográfico.* 1st ed. Sección de obras de antropología. México, DF: Fondo de Cultura Económica.

Escalante Gonzalbo, Pablo and Aban Flores-Morán. In Press. 'El Paisaje, La Mirada y La Construcción Del Espacio Colonial En El Sur Del Valle de Toluca'.

Estrada-Belli, Francisco and Heather Hurst. 2011. 'Palace Arts'. In *Mapping Latin America: A Cartographic Reader*, edited by Jordana Dym and Karl Offen, 25–8. Chicago and London: University of Chicago Press.

Fernández-Christlieb, Federico. 2015. 'Landschaft, Pueblo, and Altepetl: A Consideration of Landscape in Sixteenth-Century Central Mexico'. *Journal of Cultural Geography* 32 (3): 331–61. https://doi.org/10.1080/08873631.2015.1041307.

Florescano, Enrique. 2009. *Los Orígenes Del Poder En Mesoamérica*. México: Fondo de Cultura Económica.

Galarza, Joaquín. 1980. *Codex de Zempoala. Techialoyan E 705. Manuscrit Pictographique de Zempoala, Hidalgo, Mexique*. México: Mission Archéologique et Ethnologique Française au Mexique.

Galarza, Joaquín. 1990. *Amatl, amoxtli: el papel, el libro: los códices mesoamericanos: guía para la introducción al estudio del material pictórico indígena*. México, DF: Tava Ed.

Galarza, Joaquín. 1992. *In amoxtli in tlacatl, El libro, el hombre: Códices y vivencias*. 1st ed. Colección Códices mesoamericanos. México, DF: Tava Ed.

García Zambrano, Ángel Julián. 2006. *Pasaje Mítico y Paisaje Fundacional En Las Migraciones Mesoamericanas*. 1st ed. Cuernavaca, Morelos, México: Universidad Autónoma del Estado de Morelos, Facultad de Arquitectura.

Gibson, Charles. 1964. *The Aztecs under Spanish Rule*. Stanford, CA: Stanford University Press.

Glass, John B. and Donald Robertson. 1975. 'A Survey of Native Middle American Pictorial Manuscripts'. In *Handbook of Middle American Indians, Volumes 14 And 15: Guide to Ethnohistorical Sources, Parts Three and Four*, edited by Robert Wauchope, Howard F. Cline, Charles Gibson, and H.B. Nicholson, 3–80. Austin, TX: University of Texas Press. http://ebookcentral.proquest.com/lib/lancaster/detail.action?docID=4826328.

Gruzinski, Serge. 1991. *La colonización de lo imaginario: sociedades indígenas y occidentalización en el México español, siglos xvi-xviii*. Translated by Jorge Ferreiro. 1a reimpr. Sección de obras de historia. México: Fondo de Cultura Económica.

Gutiérrez Mendoza, Gerardo. 2012. 'Hacia Un Modelo General Para Entender La Estructura Político-Territorial Del Estado Nativo Mesoamericano (Altepetl)'. In *El Poder Compartido. Ensayos Sobre La Arqueología de Organizaciones Políticas Segmentarias y Oligárquicas*, edited by Annick Daneels and Gerardo Gutiérrez Mendoza, 27–68. México: CIESAS-COLMICH.

Harris, Trevor M. 2016. 'From PGIS to Participatory Deep Mapping and Spatial Storytelling: An Evolving Trajectory in Community Knowledge Representation in GIS'. *The Cartographic Journal* 53 (4): 318–25. https://doi.org/10.1080/00087041.2016.1243864.

Hill Boone, Elizabeth. 2010. *Relatos En Rojo y Negro. Historias Pictóricas de Aztecas y Mixtecos*. México: Fondo de Cultura Económica.

Hill Boone, Elizabeth and Walter Mignolo, eds. 1994. *Writing without Words: Alternative Literacies in Mesoamerica and the Andes*. Durham and London: Duke University Press.

Hirth, Kenneth G. 2012. 'El Altepetl y La Estructura Urbana En La Mesoamérica Prehispánica'. In *El Poder Compartido. Ensayos Sobre La Arqueología de Organizaciones Políticas Segmentarias y Oligárquicas*, edited by Annick Daneels and Gerardo Gutiérrez Mendoza, 69–98. México: CIESAS-COLMICH.

Jansen, Maarten and Gabina Aurora Pérez Jiménez. 2011. *The Mixtec Pictorial Manuscripts: Time, Agency, and Memory in Ancient Mexico*. The Early Americas: History and Culture, Vol. 1. Leiden and Boston: Brill.

Jiménez Padilla, Blanca and Samuel Villela Flores. 1999. 'Vigencia de La Territorialidad y Ritualidad En Algunos Códices Coloniales'. *Arqueología Mexicana* (38): 58–61.

Jiménez Padilla, Blanca and Samuel Villela Flores. 2003. 'Rituales y Protocolos de Posesión Territorial En Documentos Pictográficos y Títulos Del Actual Estado de Guerrero'. *Relaciones. Estudios de Historia y Sociedad* 24 (95): 95–112.

König, Viola. 2011. 'El Mapa de Teozacoalco y El Concepto de Mapamundi'. In *Pictografía y Escritura Alfabética En Oaxaca*, edited by Sebastian van Doesburg, 215–38. Oaxaca: CSEIIO/SAI/CEDELIO/FAHHO.

León-Portilla, Ascensión H. de. 1988. *Tepuztlahcuilolli, impresos en náhuatl: historia y bibliografía*. México: UNAM.

León-Portilla, Miguel. 2012. 'La Riqueza Semántica de Los Códices Mesoamericanos'. *Estudios de Cultura Náhuatl* 43 (June): 139–60.

León-Portilla, Miguel and Carmen Aguilera. 2016. *Mapa de México Tenochtitlan y sus contornos hacia 1550*. Ciudad de México: Ediciones Era, Secretaría de Cultura, El Colegio Nacional.

Lockhart, J. 1992. *The Nahuas after the Conquest*. Stanford, CA: Stanford University Press.

López-Austin, Alfredo. 1973. *Hombre-Dios: Religión y Política En El Mundo Náhuatl*. Serie de Cultura Náhuatl. Monografía 15. México: Universidad Nacional Autónoma de México, Instituto de Investigaciones Históricas.

López-Austin, Alfredo. 1997. 'El árbol cósmico en la tradición mesoamericana'. *Monografías del Real Jardín Botánico de Córdoba* (5): 85–98.

López-Austin, Alfredo. 1998. *Hombre-dios: religión y política en el mundo náhuatl*. Tercera edición. Serie de cultura náhuatl. Monografías 15. México: Universidad Nacional Autónoma de México, Instituto de Investigaciones Históricas.

López-Austin, Alfredo and Leonardo López-Luján. 2009. *Monte Sagrado-Templo Mayor. El Cerro y La Pirámide En La Tradición Religiosa Mesoamericana*. México: Instituto Nacional de Antropología e Historia/ Universidad Nacional Autónoma de México.

Manzanilla, Linda R. 2000. 'The Construction of the Underworld in Central Mexico'. In *Mesoamerica's Classic Heritage: From Teotihuacan to the Aztecs*, edited by David Carrasco, Lindsay Jones, and Scott Sessions, 87–116. Boulder: University Press of Colorado.

Manzanilla, Linda R. 2008. 'La Iconografía Del Poder En Teotihuacán'. In *Símbolos de Poder En Mesoamérica*, edited by Guilhem Olivier, 111–31. México: Instituto de Investigaciones Históricas, UNAM.

'Matrícula de Tributos'. 1522. www.wdl.org/en/item/3248/.

Molina, Fray Alonso de. 2008. *Vocabulario En Lengua Castellana y Mexicana y Mexicana y Castellana*. México: Porrúa.

Mundy, Barbara. 1996. *The Mapping of New Spain: Indigenous Cartography and the Maps of the Relaciones Geograficas*. New ed. Chicago, IL: University of Chicago Press.

Murrieta-Flores, Patricia, Mariana Favila-Vázquez, and Aban Flores-Morán. 2019. 'Spatial Humanities 3.0: Qualitative Spatial Representation and Semantic Triples as New Means of Exploration of Complex Indigenous Spatial Representations in Sixteenth Century Early Colonial Mexican Maps'. *International Journal of Humanities and Arts Computing* 13 (1–2): 53–68. https://doi.org/10.3366/ijhac.2019.0231.

Navarrete Linares, Federico. 2011. *Los orígenes de los pueblos indígenas del Valle de México: los altépetl y sus historias*. México: Universidad Nacional Autónoma de México.

Nowotny, Karl. 1961. *Tlacuilolli. Die Mexikanischen Bilderhandschriften Stil Und Inhalt*. Berlin: Verlag Gebr. Mann. Monumenta Americana.

Nuttall, Zelia. 1902. *Codex Nuttall; Facsimile of an Ancient Mexican Codex Belonging to Lord Zouche of Harynworth, England*. Cambridge, MA: Peabody Museum of American Archaeology and Ethnology. http://archive.org/details/gri_33125011146541.

Olko, Justyna. 2006. '¿Imitación, Patrimonio Pan-Regional o Distorsión Colonial? Influencia Mexica En Manuscritos Pictográficos Del Centro de México'. *Revista Española de Antropología Americana* 36 (2): 139–74.

Olko, Justyna. 2014. *Insignia of Rank in the Nahua World: From the Fifteenth to the Seventeenth Century*. Boulder: University Press of Colorado.

Oudijk, Michel R. 2002. 'La toma de posesión: un tema Mesoamericano para la legitimación del poder'. *Relaciones. Estudios de Historia y Sociedad* 23 (91): 97–131.

Peñafiel, Antonio. 1885. *Nombres Geográficos de México. Catálogo Alfabético de Los Nombres de Lugar Pertenecientes al Idioma Náhuatl.* México: Oficina Tipográfica de la Secretaría de Fomento.

Pérez-Rocha, Emma and Rafael Tena. 2000. *La Nobleza Indígena Del Centro de México Después de La Conquista.* 1st ed. Colección Obra Diversa. México, DF: Instituto Nacional de Antropología e Historia.

Portuondo, María M. 2009. *Secret Science: Spanish Cosmography and the New World.* Chicago and London: University of Chicago Press.

Proceso de Oficio. 1570–1572. 'Proceso de Oficio de La Justicia Eclesiástica. Manuel de Olvera, Corregidor de Mizquiahuala, Por Haber Entorpecido La Disposición Del Cura de Mizquiahuala Para Que Los Indios de Otros Pueblos Anexos Ocurriense a Éste a Oir Misa y Demás Prácticas Religiosas'. Archivo General de la Nación. Vol. 72, exp.7, fs. 14r–89r.

Restall, Matthew. 1997. *The Maya World: Yucatec Culture and Society, 1550–1850.* Stanford, CA: Stanford University Press.

Ridge, Mia, Don Lafreniere, and Scott Nesbit. 2013. 'Creating Deep Maps and Spatial Narratives through Design'. *International Journal of Humanities and Arts Computing* 7 (1–2): 176–89. https://doi.org/10.3366/ijhac.2013.0088.

Roberts, Les. 2016. 'Deep Mapping and Spatial Anthropology'. *Humanities* 5 (1): 5. https://doi.org/10.3390/h5010005.

Robertson, Donald. 1972. 'The Pinturas (Maps) of the Relaciones Geográficas with a Catalog'. In *Handbook of Middle American Indians, Volume 12: Guide to Ethnohistorical Sources, Part One,* edited by Howard F. Cline, 243–78. Austin, TX: University of Texas Press.

Rodríguez Cano, Laura and Alfonso Torres Rodríguez. 2001. 'Mapa Cartográfico-Histórico de Atenco-Mizquiahuala'. In *Códices Del Estado de Hidalgo/State of Hidalgo Codices,* edited by Laura Elena Sotelo Santos, Víctor Manuel Ballesteros García, and Evaristo Luvián Torres, 64–9. Pachuca, Hidalgo: Universidad Autónoma del Estado de Hidalgo.

Ruiz Medrano, Ethelia. 2001. 'En el cerro y la iglesia: la figura cosmológica atl-tepétl-oztotl'. *Relaciones* 22 (86): 141–84.

Russo, Alessandra. 2005. *El Realismo Circular: Tierras, Espacios y Paisajes de La Cartografía Indígena Novohispana Siglos XVI y XVII.* Estudios y Fuentes Del Arte En México 76. México, D.F: Universidad Nacional Autónoma de México, Instituto de Investigaciones Estéticas.

Smith, Mary Elizabeth. 1973. *Picture Writing from Ancient Southern Mexico: Mixtec Place Signs and Maps.* 1st ed. The Civilization of the American Indian Series 124. Norman: University of Oklahoma Press.

Sterpone, Osvaldo. 2001. 'Códice de Santa María Nativitas Atengo'. In *Códices Del Estado de Hidalgo/State of Hidalgo Codices,* edited by Laura Elena Sotelo Santos, Víctor Manuel Ballesteros García, and Evaristo Luvián Torres, 64–9. Pachuca, Hidalgo: Universidad Autónoma del Estado de Hidalgo.

Torquemada, Juan de. 1969. *Monarquía indiana.* México: Editorial Porrúa.

Villela Flores, Samuel. 2018. 'Altares y Ritualidad Agrícola En La Montaña de Guerrero, México'. *Dimensión Antropológica* 25 (72): 7–31.

Wright Carr, David Charles. 2009. 'Mizquiahuala En El Siglo XVI: Dominio y Resistencia En Un Pueblo Otomi'. In *Estudios de Antropología e Historia. Historia Colonial,* edited by Verenice Ramírez Calva and Francisco Jiménez Abollado, 21–54. Pachuca, Hidalgo: Universidad Autónoma del Estado de Hidalgo.

7   Deep mapping the lived world

Immersive geographies, agency, and the virtual umwelt

Trevor M. Harris

Introduction

Despite increased usage, the term deep mapping remains ill-defined, methodologically obtuse, and arguably more a vision of "what could be" rather than a mapping framework ready for use. Part of the challenge to define a methodological framework for deep mapping might relate to the various intellectual origins of deep mapping and not least mapping attempts to mirror the deep text writings of Least Heat-Moon (1991). The admonitions of Lopez (1976) to throw away the "wrong sort" of superficial "thin" maps, and to seek "less suspicious" deep maps, clearly focuses deep mapping on a narratological mapped representation of space. Pearson and Shanks (2001) suggest a second intellectual origin for deep maps in the eighteenth-century antiquarian approaches to geography and history that combine place, history, text, folklore, natural history, and hearsay. Deep maps they suggest:

> record and represent the grain and patina of a location . . . [through the] . . . juxtapositions and interpenetrations of the historical and the contemporary, the political and the poetic, the factual and the fictional, the discursive and the sensual; the conflation of oral testimony, anthology, memoir, biography, natural history and everything you might ever want to say about a place.

Small wonder then that methodological frameworks for deep mapping might be ambiguous and indistinct, though the emphasis on place is noteworthy.

In their image of deep maps, Pearson and Shanks (2001) reflect Debord's (1955) Situationist concepts and use of psychogeography and inventive strategies for mapping *dérive* (drifting) and seeking new awareness and emotional life experiences through encounters with alternative and unpredictable pathways. The concept of the lifeworld or *lebenswelt*, drawn from Husserl (1936), as a precursor to phenomenology, encapsulates the activities, experiences, and contacts that grounds the ways in which individuals experience and share the world and the world as lived. Significantly, this focus on the lifeworld and the totality of a person's direct involvement with places and environments experienced in everyday life, so reminiscent of De Certeau's (1984) spatial stories and practices of

DOI: 10.4324/9780367743840-7

everyday life, is a powerful theme in the conceptual framing of a deep map. Pearson and Shanks (2001) blurred genre reflects the coupling of thick description with a mix of scientific practice and narration to provide an integrated interdisciplinary, intertextual and creative approach to recording, writing, and illustrating the material past. Here, perhaps, in the affective ties between people and space and the combining of subjective and objective knowledge, encounter, and the use of mixed methods is where the core of deep mapping is to be found. Methodologies to achieve such a complex mix, however, are much less clearly defined.

Several more recent themes have contributed further to the conceptualization of deep mapping and have, in turn, contributed to fashion potential methodological forms. Deep mapping is one outcome of the spatial turn in the humanities and the intent to embed space and spatial concepts into the humanities (Bodenhamer et al. 2010, 2015). As such, deep mapping represents a vision and an opportunity to advance the humanities by extending theoretical and methodological constructs surrounding the role of space in human affairs. Deep mapping can also be seen as a corollary of the socio-theoretical critique of GIS and cartographic mapping arising from the GIS and Society debates of the 1990s and subsequently Critical GIS and Critical Cartography (Nyerges et al. 2011). These engagements revolve around the perceived impact of a positivist, expert, and reductionist GIS technology on the study of people, human geographies, and the environment (Harris and Weiner 1996; Harris et al. 2011). In this respect deep mapping seeks to redress the perceived deficiencies in the mapping of the humanities arising from the reductionist role of GIS and the heavy reliance on cartographic mapping and spatial analytic methods. Similarly, McLucas (Springett 2015) and Pearson and Shanks (2001) suggest that deep map representations should resist reductionist tendencies and provide a bridge for the humanities to maintain a nuanced, multivocal, open ended, and storied approach by combining qualitative measures with measured and reductionist arguments. Deep mapping draws upon methodological and technological advances that seek to overcome the binary division of qualitative and quantitative methods and provide a blended hybrid approach to humanistic studies. The need for such a hybrid approach is not unique to the humanities, for human geography and much of the social sciences share very similar challenges. Finally, deep mapping can be seen as a way of emphasizing the role of place as well as that of space in humanistic studies and to incline GIS toward a greater empathy in experiencing the lived world rather than its current heavy focus on the spatial characteristics and mapping of the physical and tangible world. Nonrepresentational theory and the affective turn reinforce this need to reexamine the dominant mapping of the material world and to look for alternative meanings and representations of space.

The hybridity of methods necessary to assess any experience of place and the encounter between the tangible physical environment and the intangible symbolic, contested, emotional, and imaginary worlds has connotations for how deep maps are generated. McLucas (Springett 2015) provides valuable insight into the character and nature of deep maps as being big, sumptuous, graphical, open ended, and digital multilayered orchestrations of media. Deep maps he suggests:

will be politicized, passionate, and partisan. They will involve negotiation and contestation over who and what is represented and how. They will give rise to debate about the documentation and portrayal of people and places . . . [and] will be unstable, fragile and temporary. They will be a conversation and not a statement.

In striving to resist reductionism, McLucas (Springett 2015) argues that deep maps will not seek the authority and objectivity of conventional cartography but rather bring together media into a new creative space. These many elements are important for the approach taken in this chapter in drawing upon the experiential power of virtual reality and immersive geographies. The proposed approach to deep mapping in this chapter shifts the representation of the lived world from a 2D cartographic form to a more intuitive 3D portrayal of reality. The platform combines GIS, 3D modeling, smell, and sound, within the powerful serious gaming engine Unity (https://unity.com/) to provide multiple ways of combining qualitative media and placial representations. The immersive qualities of the system situates the reader not as a passive observer of a fixed geography, structured and determined by a cartographic depiction, but as a reader with agency to interact, explore, and experience the media and materials: an active witness within the immersive lived world, or umvelt, of the deep map. This situatedness shifts deep mapping away from the role of being solely a holistic cornucopia of map, media, images, text, and narrative toward a platform where the reader has the agency to explore and experience a place and to form awareness, knowledge, and insight according to their positionality in the virtual scene.

Mapping and deep mapping

Part of the allure of deep mapping revolves around the perception that traditional cartographic mapping fails to capture the richness of the lived humanist world. Maps have been the primary means of communicating about space for millenia and have long served as effective representational models of reality that record and communicate the spatial components and relations of the world through a structured symbology (Guelke 1977; Kraak and Ormeling 2002; Robinson and Petchenik 1977; Slocum 2003). As Casti (2000, 11) would suggest:

The semiotic geographic map is a collection of signs and as a representational device communicates signs and messages contained in the map to an audience. Maps provide an ordered knowledge and have their own semantics, syntax, and grammar that cartographically act as a surrogate for representing the location and relationships of spatial features as they exist in reality.

Maps not only act as referents and delineate the locations of objects, features, places, and their geographic relationships but can also store vast quantities of spatial information on a single sheet or image. Map interpretation is generally achieved through good cartographic design, appropriate scale, symbology, color, classification, and content with the goal of making the map intuitive to an intended audience (Robinson 1953). Cartographic maps are the spatial equivalent to language and the written word and the widespread adoption of GIS and the geospatial web have acted to reinforce the importance of maps, invariably 2D, as essential and critical communication devices to a modern informed society. The traditional cartographic form of mapping, labeled the Map Communication Model (MCM), represents a process whereby the cartographer or map maker creates an optimally functioning map for an end user group using best design principles and practices. These maps have attained a level of authenticity, unquestioning veracity, and finality that is rarely assigned to other communication forms (Crampton 2001; Wood 1992; Ubikcan 2006).

In recent years, the MCM, as a seemingly objective map, has been heavily critiqued as the product of an expert driven, top-down process which is, in reality, a subjective representation. Maps are necessarily scaled abstractions of reality because a one-to-one representation is ineffective. As such, priorities as to what elements to emphasize and what details to suppress are important in the map design especially if the map reader is not to be overwhelmed with excessive content. This scaled abstraction and selectivity of content led Monmonier (1996, 1) to suggest that "an accurate map must tell white lies," and can knowingly or unwillingly obscure, confuse, and misrepresent. The map is thus far from an objective representation but is value laden, culturally contextualized, context-dependent, and contingent, not least in terms of the motive, power relations, and representations on the map (Crampton 2001; Crampton and Krygier 2006; Wood 1992). In particular, maps are necessarily reductionist and selective in their mapped content and tend to be dominated by the material visible features typically found on a topographic map sheet such as infrastructure, morphology, hydrography, and settlement.

The social theoretic critique of cartography and GIS questioned the mapping process as a sociopolitical discourse between those representing the knowledge being mapped and those consuming the map and its contents (Crampton 2001; Crampton and Krygier 2006; Harley 1989; Wood 1992; Ubikcan 2006). The MCM and GIS are thus perceived as positivist communication devices that privilege Boolean logic, scientific and technical authority, spatial accuracy, tangible objects and features, and reductionism, over deeper representations of space and place and particularly the lived world (Craig et al. 2002; Crampton and Krygier 2006; Harris and Weiner 1996). Through the lens of critical cartography Crampton (2001, 236) suggests an epistemological break between the positivist MCM that privileges accuracy and technical–scientific authority and connotations of objectivity, from maps that are revealed as contingent social reconstructions and where knowledge is both constructed and contested. Participatory GIS arose to counter these expert-driven, top-down representations and to incorporate bottom-up community knowledge and the many related qualitative forms by which local knowledge is recorded and represented (Craig et al. 2002). Indeed, in his seminal work *Deconstructing the Map*, Harley (1989, 3) suggests the need to deconstruct the map and to read "between the lines" and in the margins to discover the true silences and contradictions that may challenge the interpretation of mapped spatial data. In this process of deconstruction Harley questions the implied links between reality and cartographic and GIS representation, the inherent "secrecies and silences" in the map, and the social contingency of the map.

The MCM may be seen to stand in some contrast to the vision of a deep map, focused as it is on the nuanced lived world which is spatially complex, multidimensional, and sensuous, and contains the multiple realities and unfolding spatial stories and experiences that are so difficult to capture, express, and comprehend in a map representation. Spatial stories are how people understand and navigate the world and appreciate our place in it. Stories allow people to share experiences and connect to others through many forms including oral and textual narrative and visual forms and it is these stories that humanize scholarly research. Despite these criticisms, Harris (2015) has suggested referring to MCM maps as "thin maps," instead of shallow maps as the antonym to deep maps in order not to diminish their real value and contributions across a broad range of application areas. As Tuan (1974) suggests, space contextualizes place and is central in locating the geographical context within which placial themes can be addressed. A significant challenge to deep mapping, then, does not revolve solely around the concept, vision, and purpose of the deep map but around the methodology by which thin maps might be deepened. Several methods for creating deep maps have been proposed including neogeography and the geospatial web, mashups and APIs, multimedia platforms, story maps, and qualitative and participatory GIS among others (Bodenhamer et al. 2015).

Place, nonrepresentational theory, affective geographies, and the umvelt

Perhaps the essence of deep mapping lies in the goal of capturing and experiencing aspects of place, and more so than the mapping and representation of space (Harris 2015). Although somewhat simplified, space is traditionally reduced to a topological construct that digitally captures material, tangible, and geometrical properties (Dourish 2006; Giaccardi and Fogli 2008; Harrison and Dourish 1996). Place is a significant locale to which human activity and social practices and meaning are attached (Tuan 1974, 2001). In reality, as Dourish suggests, space too is a social product and practice because map makers define and select what to represent according to the practices and knowledge of their own account of space (Dourish 2006). Tuan (1974, 2001) suggests that a sense of place is not gained solely through being in a specific location but rather through some form of experience, interaction, meaning, or emotional attachment that is assigned by people to a location and place at a "lived" or "local" scale. This sense of place and attachment to a location varies across space and people.

It is suggested here that in seeking to deepen the cartographic form, nonrepresentational theory, or more appropriately theories, is a valuable concept for espousing the need to push research beyond representation and toward a focus on practice, embodiment, materiality, process, and experience (Anderson and Harrison 2010; Dewsbury 2003; Thrift 2007). In contrast to a focus on produced representations in all their forms, what Dewsbury (2003) describes as "witnessing" or knowledge production without contemplation, nonrepresentational theories focus on the embodied phenomenological experience of the lived world. In this respect nonrepresentational theory draws on the works of many, including De Certeau and his practices of everyday life. Simpson (2017) suggests that use of the term "non" presages dispatching with representations in general which troubles many geographers. Rather Simpson argues, nonrepresentational theory is intended more to reflect a different approach and a movement away from an interpretation of the meaning of representations toward a consideration of what they "do" in the unfolding of the social world. In the context of deep mapping and the desire to deepen the "thin" cartographic map, it is interesting to note Lorimer's (2005) suggestion that a more appropriate term here would be "more-than-representational." This latter term suggests not an abandonment of traditional map representation but that "the ideas proposed by non-representational theories can act as an animating supplement to existing approaches to geographic knowledge production."

In similar vein, the affective turn and the rise of affective geographies equally places greater emphasis on perception, emotion, interpretation, expectation, and feelings and the embodied experience to elicit and visualize the personal and affective meaning of people in an environment (Clough 2007; Gregg and Seigworth 2010; Stewart 2007). Cartography struggles to associate these affects based on the topological cartographic mapping and attribution form. As Giaccardi and Fogli (2008, 173) suggest, cartographic semantics are not designed to elicit and visualize "affective meaning" involving perception, interpretation, and expectations that are ascribed to a specific physical and social setting and people experience, interpret, and are affected by their mapped environmental settings.

> The cartographic semantics of current . . . [maps] show the where and when of information, but they do not visually relate that information to one's perceptions, interpretations, and expectations—they are not designed to show the personal meaning that one ascribes to specific locations.

Furthermore, they suggest that:

> being able to capture and visualize affective meaning is vital to enhance our perception of space, deepen our connections with the urban and natural environment, and stimulate reflection and discussion about the places in which we live and that we share.
>
> (ibid., 175)

Cartographic content alone then, the where and when of spatial information, is challenged by the way in which personal meaning, practices, conflicting viewpoints, and concerns interact with an environment and the objects within it. The desire here is for deep maps to mirror these richer cartographic semantics. Giaccardi and Fogli (2008, 175) suggest that weaving affective meaning and geospatial mapping into the cartographic semantics of affective geographies has the potential to provide a "visual account of how space and place relate to each other" and how personal experiences and the experiences of others reflect and inform actions and the remembering, reconstructing, and representation of space and place. Visually correlating the lived world of meaning with the cartographic world is challenging but necessary since "Only in this way can they display aspects of the environment that lie beyond our usual perception and allow multiple readings of the same territory" (ibid., 175). In seeking to counteract the reduction of reality by the MCM and GIS to a single objectified interpretation, affective maps seek to include multiple readings of the lived world and as such resonate with the powerful concept of the umvelt and deep mapping.

Drawing on nonrepresentational theory and affective mapping, it is suggested here that combining these with the bio-semiotic concept of the umwelt provides an intriguing vision and way forward in deep mapping and for communicating the human experiences of both space and place. The concept of the umwelt developed by Jakob von Uexküll (1934) proposed that within the same ecosystem different animals' sense and respond to differing environmental signals. Uexküll proposed that it is what is sensed by specific organisms in the umwelt or an environment that lies at the core of communication and signification in humans and animals. Thus, while organisms live within a network of physical objects or features that might be shared with other organisms, these organisms may share the same space but experience and occupy a different umwelt depending on their biological imperatives. Thus, while the functional components of the umwelt, such as shelter, threat, food, or features in the landscape which form the physical world of an organism, only by sensing, experiencing, and interacting with the surrounding environment do animals and people form a feedback loop that assembles an understanding of the objects of experience and that continually reshapes that umvelt based on continued interaction. While von Uexküll's ideas were later usurped to support supremacist undertakings (Agamben 2004), Tuan (1974, 2001) in his focus on the lived experiences of space and place argued that human experience is highly individualistic but that innate biological and cognitive traits are common to societies. These innate traits enable our individual experiences of the physical, the symbolic, the emotional, and the many meanings associated with place, to be shared more broadly.

Deely (2001) suggests that umvelt should be interpreted as Lebenswelt or life-world or self-centered world in that while a physical environment may be the same for all organisms living in that environment, it is "not the world in which any given species . . . actually lives out its life" (Deely 2001, 126). As a result of the distinctive biological constitution of life forms, each organism is suited to certain parts of a physical environment (Kull 1998; Sagan 2010; Sharov 2001; Skocz 2005, 2006, 2009). Significantly, the important cognitive sensing of the physical world depends on the differing sensory modalities of the multiple life forms which inhabit the physical world. To exemplify these multiple sensed realities, Deely (2001, 126) uses the example of looking out over a meadow where a poet may see a variety of colors but not see all possible colors, nor see them under changing light conditions, nor sense the meadow as do other animals such as a bee. While the physical environment may be the same, the sensing of that environment by differing life forms or differing people is not the same. This is significant for the expression of multiple realities in a deep map for it is not the measured, referenced, physical environment (the mapped space) that gives meaning to a world of objects but rather the world of objects as they are sensed and experienced by a particular organism as a place. Thus, while different organisms might physically inhabit and interact with a world of objects, this physical interaction does not in or of itself determine what will be cognitively sensed and understood as the lifeworld or umvelt. In some respects this is not dissimilar to the reality tunnel theory of representational realism that suggests that the human subconscious operates on a set of mental filters formed from experiences and belief systems that leads to individuals interpreting the same world differently (Wilson 1983). Access to reality in the umvelt is thereby mediated through human senses and non-objective factors such as experience, beliefs, conditioning, and signs (Sebeok 1976). This is significant for deep mapping and distinguishing between the cartographic mapping of the MCM and deep mapping, for while organisms might inhabit the same physical space, they sense and experience a different place which only acquires meaning as it is influenced by social context, embodiment, perception, and symbolism and when it is sensed, perceived, and experienced. While GIS and MCM may map and model the "objective" collection of objects, it is how these objects are experienced by different people as part of their own umvelt that is critical in gaining an understanding of their lived world and the multiple realities of those in the umvelten.

Immersion, virtual reality, and the virtual umvelt

People live in a multidimensional, sensuous, and spatially complex lived world—the umvelt. In this chapter deep mapping is explored using an immersive virtual reality environment, a powerful multisensory and interactive virtual spatial platform that immerses the user in a richly sensuous 3D environment capable of recreating elements of the lived world. For this reason, I add the prefix "virtual" to umvelt. Additional geosensory inputs such as sound, smell, and touch contribute substantially to enhancing the experience of that 3D immersive environment and more so than is possible through a visucentric focus on cartographic 2D maps. These experiential GIS environments enable the reductionist spatial corset of analytic space to be communicated within a complementary and collaborative setting involving the cognitive, experienced, emotional, and associative qualities of place.

Technological advances in 3D GIS, stereo-enabled virtual immersion, scenting and tactile devices, and serious gaming engines now enable powerful multisensory spatial platforms to be created. These interactive virtual platforms in the form of individualized head-mounted visualization systems such as the HTC Vive (www.vive.com) and the Oculus Rift (www.oculus.com/rift/), or the CAVE (Cave Automatic Virtual Environment) fully immerse a user in a sensuous 3D environment and enable alternative GIS representations of space and place to be generated and interrogated (Grimshaw 2014; Harris 2017; Lin and Batty 2009). The CAVE is a specially designed structure comprising three to six walls, often including a floor or a ceiling, and is generally made up of approximately 10-foot screens. Stereo images are projected onto these screens by rear-mounted projectors or bounced via a mirror onto the screen to reduce the throw distance from projector to screen. The projected images are computationally parsed into one seamless image projected across the CAVE walls (Cruz-Neira et al. 1992). Users wear 3D glasses and experience a sense of immersion within a virtually projected but very life-like sensory environment. A combination of stereo depth perception, peripheral vision capture, head tracking devices, and interactive hand controllers, contribute to creating a profound sense of immersion for the user. The holographic projection gives users an experiential sense of "presence" and of "being there" and of experiencing the projected virtual world.

Immersion is a powerful psychophysical experience that draws a user into a seemingly real but virtually rendered environment and transforms the user from a passive observer to an active participant (Harris and Hodza 2011). The visual-cognitive model draws on human pareidolia and apophenia, and the creative willingness of the mind to move seamlessly between the physical, imaginary, and virtual worlds while seemingly disengaged from the physical reality of the actual projection space itself. Experiencing immersion transports users into a multidimensional and multisensory environment in which users interact with the displayed scene in a highly interactive and dynamic manner. Not only does the virtual environment provide a visually stunning immersive insight into space and place but additional senses and emotions are also engaged to reinforce that sense of place. Sound and the sense of smell have a powerful emotional effect in triggering and evoking memories and feelings about places and other times.

Three case studies: Wheeling, Blair Mountain, and diurnal mapping

Deepening space and place—Rebecca Harding Davis's antebellum Wheeling, WV

By way of example, three immersive deep map vignettes are briefly presented to suggest ways in which immersion, serious gaming engines, and immersive technologies might contribute to deep mapping. The first vignette draws on the writings of the mid-nineteenth-century realist writer Rebecca Harding Davis about the industrializing town of Wheeling, West Virginia where she resided during much of her life (Davis 1904, 1985). Rendered 3D models of her neighborhood were reconstructed on the basis of extant buildings, historical photographs, and building footprints and using dimensions obtained from the Sanborn insurance maps of that time (Harris et al. 2016). The intent was not to attain verisimilitude of the built environment but to capture a scene representative of the geography of Wheeling at that time. The 3D buildings were embedded in a terrain model and vegetation, surface imagery, and street furniture was added to the scene. Unity, a serious game engine, was used to display the scene in the CAVE environment to create an interactive and visually stunning immersive rendering. Using particle physics, smoke belching from the dominating iron works is seen to smother the town in swirling smog-like conditions as described by Davis (Figure 7.1).

Figure 7.1 Rebecca Harding Davis's Wheeling, WV in the mid-nineteenth century.

Additional senses are engaged to reinforce the sense of place. Sounds emanating from groups of conversing bystanders in the street in their ethnic tongues, from animals roaming the streets, and from peeling church bells add to the omnipresent deep rumbling of the iron works. These holophonic and Doppler enhanced sounds are triggered as one virtually navigates through the town. Even the presence of drunken groups as described by Davis was captured with off-scene catcalls and ribald comments. Interestingly these affective ties add meaning to the streetscenes (Figure 7.2). Thus, Davis describes streets in her neighborhood where she felt unsafe and the sound of a beating heart is used to capture her emotional response to these contested places. As the reader/user navigates through virtual Wheeling the beating heart is steady but changes when the user navigates into the areas of concern identified by Davis's map of fear. As the user moves deeper into these spaces the sound of the beating heart increases in pace and diminishes as the user recedes to the margins. In this way an atavistic sound is used to emulate Davis's fight or flight reality and replicate her innate emotional response to place.

Additional geosensory inputs augment the visual and soundscape. The smells of Wheeling such as the everyday street smell of rotting detritus, thick soot, pipe tobacco, a local tannery, and a brewery permeate Davis's writings. A smell diffuser is used to create smellscapes through the virtual city whereby the scent of street garbage, livestock, cut grass and burning wood are variously inserted into the virtual scene at appropriate locations. These geosensory inputs add considerably to the place that was Wheeling as identified by Davis and communicates far more about Davis's Wheeling than the comparable and seemingly rather sterile 2D map (Ferrari 2018). The visual immersive experience of space and place as augmented by emotion, soundscapes, and smellscapes, communicates dimensions

Figure 7.2 Wheeling mid-nineteenth-century streetscene.

of depth in the experience of place and articulates the lifeworlds, places, and spatial stories of life in an industrializing town. Such a phenomenological and experiential representation of the virtual umvelt, combined with the spatial analytical power of GIS, provides a resonating means of communication about place not easily achieved through the conventional mapping of space. In this way it is possible to extend the concept of map communication beyond the physical and locational referents of space, and peer into the umvelt and the symbolic, emotional, and sensuous places of the lived world.

The contested battle for Blair Mountain

A second vignette concerns the battle for Blair Mountain in Logan County, West Virginia. Over the course of five days from late August to early September, 1921, some 10,000 armed coal miners sought to unionize the coalfields and confronted 3000 lawmen who were backed by coal mine operators (Corbin 2015; Green 2015; Savage 1986). This was the largest uprising in the United States since the Civil War and was an iconic event in the labor history of West Virginia that left a marked imprint on the psyche of the state. The confrontation was bitterly fought with an estimated one million rounds of ammunition fired and the eventual intervention of the US National Guard and US Army. This event followed the Matewan massacre of the previous year when Baldwin-Felts agents were brought in by the Stone Mountain Coal Corporation to evict unionized miners from their houses. The confrontation between miners and townspeople against the agents led to several lives being lost. Subsequent strikes, damage to property, widespread violence, and the shooting of Matewan sheriff Sid Hatfield and his deputy by

Baldwin-Felts agents on the steps of McDowell County courthouse on August 1, 1921, led to the miners' march on the state capital of Charleston, WV and subsequent march on Logan and the battle of Blair Mountain.

The perceptions, actions, memories, and experiences of those involved in the battle, as with the historical events themselves, are difficult to reconstruct. Information concerning the geographic setting of the area is even more imprecise and obscure than its history. There is a dearth of mapped information about the area or its local place names, except for coarse schematic-like renderings of events such as the miner's march toward Blair Mountain. The universally available USGS 1:24,000 topographic map is at a scale such that most place names relating to the historical record are absent. Similarly, the crowd-sourced OpenStreetMap (www.openstreetmap.org) and other web-based mapping platforms such as Google Maps are equally sparse in locational detail, often to the point of not even locating Blair Mountain itself. Satellite imagery for the area is widely available (www.wvgis.wvu.edu; Google Maps and Google Earth; Mapcarto) and display the complex topography of the area and the encircling mountaintop mining around Blair Mountain, but local knowledge is required to interpret the scenes. Google Street Scene is limited solely to pictorial views of certain state roads. The lack of a map with which to locate the battle is surprising given the impressive and remarkably precise mapping of other battlefield engagements in the United States. The spatial obscurity surrounding the geography of Blair Mountain is stark and a significant factor impacting any understanding of the actions and events that transpired there. Despite the significance of the battle of Blair Mountain in the Mine Wars of West Virginia, it is an iconic event seemingly disconnected from its geography and landscape. Indeed, much of Blair Mountain remains inaccessible because of the enforcement of trespass laws by the coal companies. Blair Mountain remains a contested landscape even today.

A poem by miner Ralph Chaplin written in 1917 about the Paint Creek and Cabin Creek strikes of 1912–13, demonstrates the intimate links between the attitude of the miners, the historical events, and the importance of the landscape of the region. The poem narrates how the miners were looking forward to spring and the return of foliage to the region that would then provide cover for an ambush:

It isn't just to see the hills beside me
Grow fresh and green with every growing thing;
I only want the leaves to come and hide me.
To cover up my vengeful wandering.
When the leaves come out, Ralph Chaplin (1917)

A 3D virtual reality model of Blair Mountain was developed within the immersive experiential CAVE, which situates the reader in the geography of the region (Figure 7.3). The reader is immersed within the deep map and geography of Blair Mountain to explore, experience, and examine a richer and deeper sense of association with the place that is Blair Mountain. Rendered 3D models of the

Figure 7.3 The virtual landscape of Blair, WV.

early-twentieth-century buildings in Blair were reconstructed based on historical photographs and the building footprints and architectural forms and dimensions were obtained from maps of that time, along with the railway, roads, and other features. The trees and foliage recounted by Chaplin are present in the virtual scene and the reader can position themselves at any vantage point in the scene and review the tableaux around them. The spatial stories of the events surrounding the battle including its history, folklore, and oral narratives from both sides combines with the physical environment to virtually recreate the umvelt and affective ties that link the lived experiences and multiple realities of those in the conflict at the time to the site. Without even a mark on a cartographic map to denote the extent of the battle, such an immersive representation begins the process of deepening the exploration of Blair. It is central that deep maps accommodate diverse views, preserve contradictions, and reflect the varied viewpoints of the participants. Rather than focus on the accuracy of the map, deep maps emphasize both space and place. Immersion and the sense of "presence" enables Blair and the battlefield to be experienced as if one were physically present in the landscape.

Previous 2D line-of-sight studies, for example, to determine what could be seen by either party at the battle are replaced through the ability to be in the virtual landscape and personally experience the viewshed from any location or vantage point (Nida 2014). The integration of stories told about peoples' lived experiences of the events surrounding the battle might include oral histories, life histories, biographies, and narratives to contribute to the nuanced dimensional richness of Blair and Blair Mountain. In this study, geographical fidelity was not a goal but immersive geography does allow a unique opportunity to peer through foliage like that of Chaplin's poem to see the landscape of early-twentieth-century Blair Mountain laid out before us (Figure 7.4). Furthermore,

Figure 7.4 Virtual Blair, WV.

the contestation over Blair Mountain continues today. In 2009 Blair Mountain was chosen for placement on the National Register of Historic Places but was challenged by a law firm representing multiple coal companies and subsequently removed from the register. Two of the largest coal companies hold permits for surface mines on Blair Mountain. These conflicting perspectives and narratives being played out on Blair Mountain are a powerful reason for deep maps to resist reductionism and maintain the multi-vocality surrounding events and places. Bauch's (this volume) use of "informed speculation" and his argument for moving beyond design neutrality and the mechanical positioning of objects on a cartographic plane and toward argumentation and conversation is particularly relevant here.

Changing light and environmental conditions

A third study dwells further on a significant limitation of cartographic mapping and the inability to display differing lighting conditions or weather or environmental conditions such as smog. The latter point was developed as part of the Wheeling study when, according to Davis (1904), swirling smog from the iron mills and home fires kept the town in semipermanent darkness. Circadian rhythms dominate human life and everyday behavior and activities in space, driven as we are by our biological clock and impelled by zeitgebers such as daylight which govern our everyday activities. Almost all aspects of spatial activity, from the daily commute, travel, energy usage, recreation, telecommunication demands, to the incidence of crime, our use of geographical space is tied to the cycles of natural light. Activities and walking routes undertaken under daylight conditions, for example, walking down certain streets or alleyways, are approached differently, depending on the individual, when nighttime conditions or poorly lit streetlights are the only light available. Anthropogenic lighting, especially in the built environment, is a major factor ameliorating the impact of darkness or poor weather conditions on human activity and any disruption to this artificial lighting, such as occurs during blackouts, has major impacts on local populations. Yet despite the centrality of differing natural, anthropogenic, and atmospheric conditions, maps are poorly equipped to display these conditions. Light rarely permeates GIS as an explanatory variable and GIS mapping appears immune to human circadian rhythms. Rather maps and GIS reflect a hypothetical setting of midday with perfect lighting and visibility in all directions even though our everyday experiences indicate that constantly changing light and atmospheric conditions have a major impact on human activity patterns.

In this example, natural lighting conditions under daylight or moonlight conditions can be displayed in interactive scenes in the 3D immersive CAVE (Kaufman 2014) (Figure 7.5). The time of day clearly influences not just light conditions, but the effect of shadows cast. The effect of differing anthropogenic lighting conditions can be examined and areas of less-than-optimal lighting can be identified. The lighting infrastructure can affect the light quality and vary by type of street bulb or color. The influence of ambient lighting from street stores, or residential properties can be examined. The placement of emergency telephones on campus can be modeled and examined and the impact of light pollution explored (Figure 7.6). Haze and fog, and smog can also be displayed as dynamic aspects of a scene based on the particle physics capability of serious gaming engines. Such a combination of lighting scenarios is important to many geographical applications ranging from geodesign, the identification of poorly lit areas and the placement of differing artificial lighting, to exploring the complex and nuanced issues of space and place in human behavior. More specifically, light and weather conditions are important to those in the lived world of the umvelt where responses to light can differ based on gender, age, or ethnicity. The immersive virtual world thus provides far greater capability than a mapped representation.

Figure 7.5 Nighttime lighting in a deep map.

Figure 7.6 Nighttime and the placement of emergency telephones.

Conclusion

Cartographic mapping and GIS strive to capture, record, and communicate the spatial components of the umvelt or lived world through pseudo-objective representations of space, primarily in the form of 2D mapping. Such representations are powerful and yet have significant limitations and implications for representing and communicating knowledge about the lived world for they are based on the allure of reductionist technologies. Nonrepresentational theory and affective mapping propose that mapped knowledge is incomplete without the meaning, associations, memories, and emotions that make these spaces and places significant. The concept of the umvelt suggests that the mapped world of objects only makes sense when it is experienced by differing organisms or people and that these multiple realities affect the relationships between objects and those in that place. Grappling with multiple realities and creating meaning from non-reductionist human complexity through deep mapping is to unravel aspects of place as well as space. These innovative technologies allow users to be immersed in the lived world and to pursue spatial stories related to individual and community goals, perspectives, and aspirations. The visual immersive experience of space and place augmented by emotion, soundscapes, and smellscapes communicates the experience of place and articulates the lifeworlds, places, and spatial stories of life in the region. Such a phenomenological and experiential representation of place focuses on how people experience and emotionally respond to an environment and the multiple encounters and stories within that environment. The umvelt comprises multiple unfolding spatial stories and experiences that are sensual and emotional, and represent associations based on space and the perception and the experience of place. These deep mappings are so difficult to capture, map, express, represent, or comprehend.

The use of 3D modelling, sensuous GIS, serious gaming engines, and immersive environments provide a powerful platform within which to create deep maps and to interrogate and experience them. Not only does the use of virtual immersive geography enable the deep map of the virtual umvelt and the lived world to be recreated and explored but it also empowers the reader to negotiate through the deep map, to explore and take differing pathways or drift, to attain certain viewpoints, to experience a sensuous, emotional, and multidimensional world, to interpret their experiences of the virtual umvelt, and to combine materials and conditions in differing ways to seek insight to the spatial stories to be told. The MCM and conventional cartographic modeling struggle to represent the lived world. To date, much attention has been given to methodological frameworks capable of collating the breadth of media and disparate content necessary to be called a deep map. Little attention has been given, however, to how the reader might be empowered and given agency to navigate and explore the space and material to create new insight. This theme of assigning authority, agency, and ethicacy to the user is present in many of the chapters in this volume and in deep mapping. Such agency in deep mapping wrests control over the spatial narrative away from the originator and transitions the map from a statement-like monolog to a dialog. Ultimately such a shared experience of space and place, where the reader is an active participant in the exploration of the deep map, could potentially contribute to a shared world of insightful spatial storytelling and understanding.

Acknowledgments

Acknowledgment and grateful thanks is hereby made to Christabel Devadoss, Fang Fang, Aaron Ferrari, Trevor Harris, and Deborah Kirk for Figures 7.1 and 7.2; to Jeffrey Cazvenas, Nicholas Peretti, Jonathan Suite, Alex Stout, Jared Wolak, and Amber Williams for contributions to Figures 7.3 and 7.4; and to Andrew Kaufman for Figures 7.5 and 7.6.

References

Agamben, G. 2004. *The Open: Man and Animal*. Translated by K. Attell. Stanford, CA: Stanford University Press.
Anderson, B. and P. Harrison, eds. 2010. *Taking-Place: Non-Representational Theories and Geography*. Burlington, VT: Ashgate.
Bodenhamer, J.D., J. Corrigan, and T.M. Harris, eds. 2010. *The Spatial Humanities: GIS and the Future of Humanities Scholarship*. Bloomington, IN: Indiana University Press.
Bodenhamer, J.D., J. Corrigan, and T.M. Harris, eds. 2015. *Deep Mapping and Spatial Narratives*. Bloomington, IN: Indiana University Press.
Casti, E. 2000. *Reality as Representation*. Bergamo, Italy: Bergamo University Press.
Chaplin, R. 1917. 'When the Leaves Come Out'. Accessed September 2020. https://en.wikisource.org/wiki/When_the_Leaves_Come_Out/When_the_Leaves_Come_Out.
Clough, P.T., ed. 2007. *The Affective Turn: Theorizing the Social*. Durham: Duke University Press.
Corbin, D.A. 2015. *Life, Work, and Rebellion in the Coal Fields: The Southern West Virginia Miners, 1880–1922*. Morgantown: West Virginia University Press.
Craig, W., T.M. Harris, and D. Weiner, eds. 2002. *Community Participation and Geographic Information Systems*. London: Taylor and Francis.

Crampton, J.W. 2001. 'Maps as Social Constructions: Power, Communication, and Visualization'. *Progress in Human Geography* 25: 235–52.

Crampton, J.W. and J. Krygier. 2006. 'An Introduction to Critical Cartography'. *ACME: An International Journal for Critical Geosciences* 4 (1): 11–33.

Cruz-Neira, C., D.J. Sandin, T.A. DeFanti, R.V. Kenyon, and J.C. Hart. 1992. 'The CAVE: Audio Visual Experience Automatic Virtual Environment'. *Communications of the ACM* 35 (6): 64–72.

Davis, R.H. 1904. *Bits of Gossip*. Cambridge: Houghton, Mifflin and Company.

Davis, R.H. 1985. *Life in the Iron-Mills and Other Stories*. New York: Feminist Press, the City University of New York.

Debord, G. 1955. 'Introduction to a Critique of Urban Geography'. In *Situationist International: Anthology*, edited by K. Knabb. Berkeley: Bureau of Public Secrets. Reprinted in 2006.

De Certeau, M. 1984. *The Practice of Everyday Life*. Translated by S. Rendall. Berkeley: University of California Press.

Deely, J. 2001. 'Umvelt'. *Semiotica* 134 (1/4): 125–35.

Dewsbury, J.D. 2003. 'Witnessing Space: "Knowledge without Contemplation"'. *Environment and Planning A* 35: 1907–32.

Dourish, P. 2006. 'Re-Space-ing Place: "Place" and "Space" Ten Years on'. *Proceedings CSCW'06*, ACM Press, New York, 299–308.

Ferrari, A. 2018. 'Exploring the Geosensory in Geography: Examining Olfaction and Geo-Virtual Immersion as Contributors to a Sense of Place and Embodiment'. Unpublished MA Thesis, West Virginia University, Morgantown, West Virginia.

Giaccardi, E. and D. Fogli. 2008. 'Affective Geographies: Toward a Richer Cartographic Semantics for the Geospatial Web'. AVI, *Proceedings of the Working Conference on Advanced Visual Interfaces*, Napoli, Italy, 173–80.

Gregg, M. and G.J. Seigworth, eds. 2010. *The Affect Theory Reader*. Durham: Duke University Press.

Green, J.R. 2015. *The Devil Is Here in These Hills: West Virginia's Coal Miners and Their Battle for Freedom*. New York: Grove Atlantic.

Grimshaw, M., ed. 2014. *The Oxford Handbook of Virtuality*. Oxford: Oxford University Press.

Guelke, L., ed. 1977. *The Nature of Cartographic Communication*. Cartographica Monograph, 19. Toronto, ON: B.V. Gutsell.

Harley, J.B. 1989. 'Deconstructing the Map'. *Cartographica* 26: 1–20.

Harris, T.M. 2015. 'Deep Geography: Deep Mapping: Spatial Storytelling and a Sense of Place'. In *Deep Mapping and Spatial Narratives*, edited by D. Bodenhamer, J. Corrigan, and T.M. Harris, 28–53. Bloomington, IN: Indiana University Press.

Harris, T.M. 2017. 'Deep Mapping and Sensual Immersive Geographies'. In *The International Encyclopedia of Geography: People, the Earth, Environment, and Technology*, edited by D. Richardson, N. Castree, M.F. Goodchild, A. Kobayshi, W. Liu, and R.A. Marston. Hoboken, NJ: Wiley.

Harris, T.M. and P. Hodza. 2011. 'Geocollaborative Soil Boundary Mapping in an Experiential GIS Environment'. *Cartography and Geographic Information Science* 38 (1): 20–35.

Harris, T.M., H.F. Lafone, and D. Bonenberger. 2016. 'Beyond Mapping Text in Space to Experiencing Text in Place: Exploring Literary Virtual Geographies'. In *Literary Mapping in the Digital Age*, edited by D. Cooper, C. Donaldson, and P. Murrieta-Flores, 221–39. Farnham: Ashgate.

Harris, T.M., L.J. Rouse, and S. Bergeron. 2011. 'Humanities GIS: Adding Place, Spatial Storytelling and Immersive Visualization into the Humanities'. In *Geohumanities: Art,*

History, Text at the Edge of Place, edited by M. Dear, J. Ketchum, S. Luria, and D. Richardson, 226–40. Abingdon, UK: Routledge.

Harris, T.M. and D. Weiner. 1996. 'GIS and Society: The Social Implications of How People, Space and Environment Are Represented in GIS'. Scientific Report for NCGIA Initiative #19 Specialist Meeting, University of California at Santa Barbara, Santa Barbara, CA, November.

Harrison, S. and P. Dourish. 1996. 'Re-Place-ing Space: The Roles of Place and Space in Collaborative Systems'. *Proceedings CSCW'96*, ACM Press, New York, 67–76.

Husserl, E. 1936. *The Crisis of the European Sciences and Transcendental Phenomenology: An Introduction to Phenomenological Philosophy*. Translated by D. Carr. Northwestern University Studies in Phenomenology and Existential Philosophy. Evanston, IL: Northwestern University Press, 1970.

Kaufman, A. 2014. 'Shedding Light on GIS: A 3D Immersive Approach to Urban Lightscape Integration into GIS'. Unpublished MA Thesis, West Virginia University, Morgantown, West Virginia.

Kraak, M.-J. and F. Ormeling. 2002. *Cartography: Visualization of Spatial Data*. Hoboken, NJ: Prentice Hall.

Kull, K. 1998. 'On Semiosis, Umwelt, and Semiosphere'. *Semiotica* 120 (3/4): 299–310.

Least Heat-Moon, W. 1991. *PrairyErth: A Deep Map*. Boston: Houghton Mifflin Company.

Lin, H. and M. Batty, eds. 2009. *Virtual Geographic Environments*. Beijing: Science Press.

Lopez, B.H. 1976. *Desert Notes: Reflections in the Eye of a Raven*. Kansas City, Missouri: Andrews McMeel Publishers.

Lorimer, H. 2005. 'Cultural Geography: The Busyness of Being "More-Than-Representational"'. *Progress in Human Geography* 29 (1): 83–94.

Monmonier, M. 1996. *How to Lie with Maps*. 2nd ed. Chicago, IL: The University of Chicago Press.

Nida, B.D. 2014. 'Patterns of Liberation: Archaeological and Spatial Analysis at the Blair Mountain Battlefield'. Unpublished PhD Dissertation, University of California, Berkeley, 1921.

Nyerges, T., H. Couclelis, and R. McMaster, eds. 2011. *The SAGE Handbook of GIS and Society*. London: SAGE.

Pearson, M. and M. Shanks. 2001. *Theater/Archaeology*. Abingdon, UK: Routledge.

Robinson, A.H. 1953. *Elements of Cartography*. New York: Wiley.

Robinson, A.H. and B. Petchenik. 1977. 'The Map as Communication System'. *Cartographica* 19: 92–110.

Sagan, D. 2010. 'Umvelt after Uexküll'. In Uexküll, Jakob von. *A Foray into the Worlds of Animals and Humans: With a Theory of Meaning*. Translated by Joseph D. O'Neil, 1–35. Minneapolis: University of Minnesota Press, 1934.

Savage, L. 1986. *Thunder in the Mountains: The West Virginia Mine War, 1920–21*. Pittsburgh: University of Pittsburgh Press.

Sebeok, T.A. 1976. *Contributions to the Doctrine of Signs*. Bloomington, IN: Indiana University Press.

Sharov, A. 2001. 'Umvelt Theory and Pragmatism'. *Semiotica* 2001 (134): 211–28.

Simpson, P. 2017. 'Non-Representational Theory'. Accessed September 2020. www.oxfordbibliographies.com/view/document/obo-9780199874002/obo-9780199874002-0117.xml.

Skocz, D.E. 2005. 'Wilderness Management and Geospatial Technology: A View from the Black Forest'. *Environmental Philosophy* 2 (2): 53–60.

Skocz, D.E. 2006. 'Ecology, Technology, and Wilderness Management: A Clash of Eco-Spatial Paradigms in Ecoscapes'. In *Ecoscapes: Geographical Patternings of Relations*, edited by G. Backhaus and J. Murung, 115–38. New York: Lexington Books.

Skocz, D.E. 2009. 'Environmental Management in the Age of the World Picture'. In *Heidegger and the Earth: Essays in Environmental Philosophy*, edited by L. McWhorter and G. Stenstad, 123–43. Toronto, ON: University of Toronto Press.

Slocum, T. 2003. *Thematic Cartography and Geographic Visualization*. Hoboken, NJ: Prentice Hall.

Springett, S. 2015. 'Going Deeper or Flatter: Connecting Deep Mapping, Flat Ontologies and the Democratizing of Knowledge'. *Humanities* 4: 623–36.

Stewart, K. 2007. *Ordinary Effects*. Durham: Durham University Press.

Thrift, N. 2007. *Non-Representational Theory: Space, Politics, Affect*. London: Routledge.

Tuan, Yi Fu. 1974. *Topophilia: A Study of Environmental Perception, Attitudes and Values*. New York: Columbia University Press.

Tuan, Yi-Fu. 2001. *Space and Place: The Perspective of Experience*. Minneapolis: University of Minnesota Press.

Ubikcan. 2006. 'The Map Communication Model and Critical Cartography'. Accessed September 2020. http://ubikcan.blogspot.com/2006/08/map-communication-model-and-critical.html.

Uexküll, J. von. 1934. *A Foray into the Worlds of Animals and Humans: With a Theory of Meaning*. Translated by Joseph D. O'Neil. Minneapolis: University of Minnesota Press, 2010.

Wilson, R.A. 1983. *Prometheus Rising*. Grand Junction, CO: Hilaritas Press.

Wood, D. 1992. *The Power of Maps*. New York: Guilford Press.

8 Navigating through narrative

Stephen Robertson and Lincoln A. Mullen

The idea of the deep map, as propounded by scholars in the humanities and geography over the last several years, aims to bring humanistic concerns to mapping. The adjective deep in this case expresses the concept of multiplicity. The aspiration of deep maps as captured in its founding definitions is to capture a multiplicity of sources used, of perspectives represented, of experiences captured, of interfaces used within the framework of the deep map.[1] This variety of inputs to the deep map is central to the concept. It distinguishes the deep map from other kinds of digital humanities approaches, especially computational methods, which too often rely on a single kind of source such as documents in text analysis or photographs in image analysis, in contrast to the more traditional approach of humanities scholars who are willing to use any and every source in the pursuit of understanding. The multiplicity of sources is an important part of what makes deep mapping a truly humanistic approach.

Even more central to the idea of a deep map, however, is the concept that a deep map not only should reflect multiple perspectives from its sources but also should enable multiple meanings to be made by its users. In contrast to current practices of humanistic argumentation as expressed in the monograph and journal article, or for that matter in thematic maps that offer an argument or conclusion about the object of study, the deep map promotes a multiplicity of meanings as part of its structural logic, embedding that orientation in its shape. Most humanistic maps are created by authors to advance certain scholarly claims. The creators of those maps have engaged in the task of selecting and arranging materials (the defining aspect of a historian's work) toward that end. But a deep map aims to take the broadest possible approach to selection, and it arranges the materials only spatially. A deep map invites the user of the map to craft his or her own experience through the multiplicity of sources made available. The path that one takes through the deep map constitutes the meaning that one makes from it. A deep map thus turns the task of authorship over to the user, at least in part. The creator of the deep map not only uses it as a research tool for his or her own purposes but also turns over the deep map to its users, who share in the process of authorship to the extent that they create their own pathways through the material. This freedom of source material on the one hand and arrangement on the other is felt to best be suited to the spirit of humanistic scholarship, where even arguments are

DOI: 10.4324/9780367743840-8

successful not primarily if they close down a question definitively but if they open up that question to new inquiry.

This impulse toward a multiplicity of sources and a multiplicity of authors is not unique to the deep map. As a field, the spatial humanities are for the most part separate from digital history, and conversations in the two fields have mostly taken place apart from one another.[2] Deep mapping nevertheless bears an essential similarity to the digital humanities, not simply because the two fields use technology, but because of the new possibilities for argument and narrative that using technology enables. A deep map is in many respects like a digital collection of primary sources. Some of the earliest visions for digital history work, such as Robert Darnton's call for a pyramid of sources leading to interpretations, understood the potential of new media to be providing access to sources so that not only the scholar but also the audience could create interpretations.[3] Likewise, the concerns of deep mapping are analogous to those of the first and second wave of digital history, the first wave of which focused on collecting and publishing materials online, and the second which focused on allowing users to have interactivity with those sources.[4] The seminal *Valley of the Shadow* project created by William Thomas III and Edward Ayers is not primarily a map, but its collection of materials arranged spatially from a Virginia slaveholding county and a Pennsylvania free county essentially aims to accomplish the same purpose of a deep map: to explore the meaning of a place through a multiplicity of sources and perspectives, which they call a "prismatic" approach.[5] Anne Sarah Rubin—also a key collaborator on the *Valley of the Shadow* project—created a layered map of memories of Sherman's march across *Georgia for Mapping Memory: Sherman's March and America*.[6] Both forms—the deep map and the digital collection—seek to gather a variety of primary materials, and both express their underlying argument only implicitly, seeking instead to permit users to navigate through their materials to make their own meaning out of them.[7]

Given the similarity of their aims, both fields are facing a question about the place of argument or narrative. We are now at a point where we want more from mapping than a means of exploring sources. Digital historians are facing calls for digital history to make arguments and are looking to expand their engagement with the discipline in order to extend the use of digital tools and to integrate them with conventional methods.[8] Likewise, it appears that the audiences for humanities mapping do not necessarily want to explore a map to find an argument or construct a narrative, or are not equipped to do so. Without a place to begin, and a pathway to follow, they spend little time engaging with digital maps.[9]

The conceptualization of deep maps has followed the same general trend as the digital humanities. For almost a decade, spatial humanists have moved from historical GIS, to spatial humanities, to deep maps, and finally to spatial narratives. That discussion has remained largely conceptual and methodological: as Trevor Harris argues, "Deep mapping, spatial storytelling, and spatial narratives are incomplete terms struggling to capture and imbue meaning to abstract thoughts of a more profound, insightful, reflexive, multimedia, perhaps quixotic representation of humanistic space than currently prevails."[10] The focus of GIS on spatial connections, patterns and relationships, and its difficulty in dealing with time, has frustrated efforts to use it to answer humanities questions concerned with process, structure, and event in the context of place. Deep maps are the first part of imagining something different: a map of place not space, a shift in perspective from an emphasis on measured space to the connotations, associations, and meanings that people attach to spaces. A map of place involves a heterogeneity of different kinds of sources, multiple viewpoints and competing perspectives, and spatial data of different degrees of ambiguity, uncertainty, and imprecision. That plural character means deep maps must be explored; by design, they do not—on their face—offer any kind of metanarrative. The act of creating a deep map is, however, implicitly argumentative: scholars much select materials to include and arrange them on the deep map, even if that selection and arrangement is done with a lighter touch than in, for example, an explicitly argumentative form of scholarship.

However, deep maps are not conceived as an end in themselves, but as the basis for constructing a spatial narrative within which an argument is embedded. The concept of spatial narrative addresses the same need for maps to be more than something to be explored raised by calls for digital historians to make arguments. Deep maps provide the context for spatial narratives, which are specific pathways through the map designed to tell a story. Conceiving spatial narratives as pathways highlights their visual character; they appear on a map. In that form, they require a specified point of departure, a series of pre-selected locations (and evidence) and an endpoint. However, there is tension between the linearity inherent in the notion of a pathway and the ability to capture complexity that maps offer: a pathway forces a narrow focus, allowing only one location for each point in time, at odds with the ability of a map to capture "the simultaneity of lived existence," to integrate and visualize multiple events happening at the same time.[11] As David Bodenhamer has put it about mapping, "its core strength is an ability to integrate, analyze, and make visual a vast array of data from different formats, all by virtue of their shared geography."[12] Maps are a way of bringing together many sources by arranging them on the spatial grid. This arrangement brings together many items which share a proximity in space and thus a mutually constitutive of a place.[13]

A spatial narrative must be more than a pathway if it is to tell a story. A sequence of points can readily convey what, where, and when; a narrative is concerned also to explain why rather than just chronicling events. As Bodenhamer has pointed out, questions of why are perhaps the primary question of the humanities.[14] How much explanation visualization can provide, how much text needs to be included, and how the visual and textual are located and balanced, are key issues in designing spatial narratives. So, where the concept of spatial narrative as constructed from deep maps begins with the map as a visual means of understanding, a narrative map by contrast begins with the textual. The point of a narrative map is not to display data but to provide an explicit visual counterpart to the implicit spatial underpinnings of a narrative or argument. The truly narrative map, as opposed to the merely linear pathway through a map, must have the same ability as a deep map to display multiplicity, even as it imposes an additional level of narrative arrangement above the mere arrangement in geographic space. The spatial narrative assumes that proximity in space, just like ordered chronology in time, is related to causation. Causation could be understood in a specific sense, in that the spatial situation caused events to happen the way that they did. Or causation could be understood in a more generic sense, in that the meaning of the place arose out of the interplay of the various sources.

The spatial narrative may in fact be the key to achieving the multiplicity which is the aim of a deep map. Maps made within well recognized genres may speak for themselves. Thematic maps, for instance, which represent single types of information within an implicitly argumentative frame, are easily understood.[15] But it seems unlikely that deep maps can speak for themselves. They could be, at worst, an amalgamation of sources, arranged spatially for sake of space, but having no other principle to guide them. Including many materials for the purpose of representing many perspectives may have the effect of representing none of them. That a multiplicity of voices, of perspectives, and of sources is mute in the midst of its chaos. The aim of the deep map is not to let the past speak for itself, because it cannot. The aim of the deep map should be to translate for the past so that it can speak. What is needed is a way of navigating through that chaos to craft meaningful experiences or experience the meaning of the past. And that act of translation, like much history, can come through narrative.

Narrative is a powerful means, perhaps the means, of making meaning out of chaos. It partakes of the fundamental acts of historical thinking. It selects the materials that will comprise the narrative. It arranges them to make meaning out of them. In many instances it periodizes them as a part of telling that story, thus contributing to the shaping of historical knowledge. It connects disparate elements and disconnects the merely accidental correspondence. A "thick description," after all, is built up not by layering on ever more descriptions of what happened, but by explaining the significance of what happened.

The concept of deep mapping might seem to be in opposition to this idea of a narrative. A narrative, after all, presumes an author and an author presumes an authority. By selecting and arranging materials, a narrative closes off some stories that could be told. We think, however, that a narrative need not close off the potential for multiple stories or narrative. After all the idea that historians should explain "how things actually were" implies that they should capture the meaning behind events, but also that there are other competing ways of telling the story. The selection and arrangement of sources necessarily imply that there are alternative selections and arrangements. Narrative maps and deep maps are not in opposition to one another; spatial narrative represents a way to navigate the deep map. Through the practice of generous affordances, it is conceivable that a narrative deep map could allow the user to craft their own stories. Even if not, the presence of other elements on the map and around the narrative allows the user to suggest revisions and ask questions even if those alternative narratives are not reified within the interface itself.

We propose, then, an understanding of the relationship between deep maps and narrative in which a user navigates a deep map through a narrative created by the author. To build on the theoretical justifications detailed earlier and in other authors' writings on spatial narratives, we would like to provide a concrete demonstration of the possibility. And so, after reviewing other platforms for combining maps and narratives, we will turn to a prototype narrative map Robertson created using the Neatline platform to explain the 1935 Harlem riot. This spatial narrative is based on the deep maps created as part of the *Digital Harlem* and *Year of the Riot* projects, but it extends them as a way of navigating through the maps to explain the riot.[16]

The combination of narrative and maps is hardly new. But we wish to distinguish our approach from a popular genre of tools called "story maps." StoryMapJS is a freely available software product created by the Knight Lab at Northwestern University.[17] This tool features the ability to string together a set of points on a map, annotating each with text, images, video, and hyperlinks. StoryMapJS forces authors who use it to create a linear narrative. A story map is much more linear even than narrative prose since the interface forces the user to march through a series of points. Even more importantly, the interface focuses on a single point at a time. While the user can see the previous and subsequent points if they are close enough in space, that spatial context serves primarily to visualize the strict linearity of the narrative. The story maps thus do not take advantage of the key feature of spatial work: the ability to organize multiple types of sources and to see them in spatial context. StoryMapJS as a model of mapping and prose thus combines the most limiting aspect of prose with only the barest kind of spatial contextualization.

ESRI's Story Map is a platform for a different kind of spatial narrative: not a pathway through a map, but a series of maps, accompanied by a narrative. As a white paper published by ESRI spells out, text "plays a supporting role, with the map or series of maps taking center stage." Their logic is simple (though probably incorrect): "internet and mobile users aren't willing to read very much." SMJournal, one of the series of templates that make up Story Map, displays text in a narrow scrolling panel alongside the map, rather than linked to points on a map, as in StoryMapJS. The text panel is only about 6 words wide, making it difficult to tell a complex story. There is some scope for interaction between text and map: links in the text can toggle layers on and off, zoom the map, and display a pop-up of a place. Further interactive elements have been omitted "so that users don't have to figure out several functions."[18]

A fully realized example of a historical spatial narrative using this platform exists: *Mapping Occupation: Force, Freedom and the Army in Reconstruction*, a project led by Gregory Downs and Scott Nesbit (Figure 8.1). It documents the locations of the US army in the south from 1865 to 1880, and the reach of their presence, as "zones of occupation" (areas the army could travel to in order to address offenses) and "zones of access" (areas within which freed people could travel to bring complaints). The narrative "guid[es] the user through key stages in the spatial history of the army in Reconstruction," with ten sections of text, each accompanied by a map, most of which are interactive to the extent you can click on the points.[19] The narrative incorporates some images, which appear as only

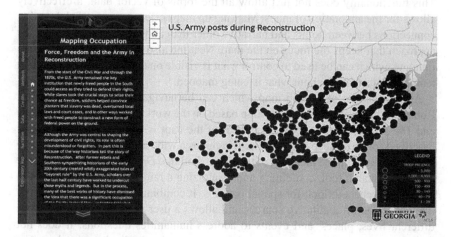

Figure 8.1 Mapping Occupation is an example of a spatial narrative using ESRI's Story Map.

slightly larger than a thumbnail due to the constraints of the panel. The maps are essentially illustrative; while at points there are links in the narrative to toggle layers because the links are in the narrative they do not encourage interaction with the map itself.

Our point in describing StoryMapJS and ESRI's Story Maps, as well as Downs and Nesbit's use of the latter, is not primarily to critique that project but to identify the underlying form of the spatial narratives enabled by those software platforms. The ideal of a spatial narrative could be more nearly approached, we suggest, by using a different software package with different affordances. Neatline, created by Scholars Lab at the University of Virginia, in collaboration with staff at RRCHNM, offers a platform with more potential for a spatial narrative of the riot than either Story Map platform.[20] Neatline adds spatial functionality to the Omeka platform, and its ability to combine rich descriptions of sources through metadata and layer them on a map makes it a generalized platform for creating a deep map. Omeka permits users to create items which correspond to many kinds of item types. Those might be various kinds of primary sources such as letters, archival records, videos, photos, datasets, or the like. But they might just as well be more abstract types to be represented on a deep map, such as events, people, or places. Each item can have metadata associated with it appropriate to the type, as well as actual data in the form of attached files, photographs, transcribed texts, or the like. The Neatline plugin then allows the user to represent those points on a map. An author can not only pick from the standard points or lines or polygons markers on a map but also has the full power of imported SVG images and web standards to represent data with more complexity. The Neatline interface gives most of the space on the screen to a map, with a sidebar not made up of text, but of a series of "waypoints," links that open windows on the map that can contain text or media.

This functionality does not just allow all the forms of vector data; it effectively allows for the annotation of maps. Marking up maps offers a means of elaborating connections between points and narrative other than adding or appending text, an approach that keeps the focus on the map. Neatline also offers a basic interactive timeline, which can be used to control what is visible on the map as well as adding temporal information about what is being mapped.

As an example of a deep map which can be navigated through a narrative, Robertson has created a Neatline map based on the work done for *Digital Harlem* and *The Year of the Riot* projects.[21] Year of the Riot is like the most commonly described forms of deep maps in that it includes a wide array of sources on a map in order to build up the meaning of the place. And like *Digital Harlem*, *Year of the Riot* is a research tool, combining and visualizing a range of different sources to allow researchers to look for spatial patterns. It combines and organizes a disparate collection of qualitative sources and allows a user to explore and juxtapose different lives, places, and events to address humanities questions. It does not offer any interpretation, argument, or narrative; it simply displays the data. The map of the events of the riot in *Year of the Riot* does include an argument in the sense that it is shaped by choices about what to include in the database, and how to categorize and present that information.[22] But that kind of argument is very different from providing a structured way into and through the information. Thanks to the timeline, the map shows what, where, and when; you can use the timeline to move sequentially through the events on the map. But *Year of the Riot* does not address the connections between those events and does not explain why they occurred and what they mean that is the stuff of historical narratives and arguments.

The *Year of the Riot* map points to the limits of these platforms. Riots are complex and multifaceted events (Figure 8.2). While they usually have an identifiable origin, they spiral out in multiple different directions, and change shape over the course of time. Riots are not linear, so are not easily rendered in narrative or with tools that focus on sequencing points or maps. A pathway won't tell the story in this map, nor will linear text in a sidebar, alongside a map. After offering a spatial narrative of the riot based on the map from *Year of the Riot*, we then prototype to what extent that narrative can be presented using Neatline.[23] Despite the limits of the prototype, we hope that this case study can move forward discussion of spatial narrative.

The map of the events of the riot contains elements at odds with existing accounts. It is crucial to note that the map is not a complete picture of the riot or even of the evidence of specific events that took place that night. It includes at most one-third of the more than 300 properties damaged during the riot, and only 68% of the 132 men and women arrested during the riot. A greater proportion of the 80 individuals assaulted, killed, or seriously injured, 83%, appear on the map. The map is also not an unmediated presentation of this evidence. To facilitate the exploration of patterns, Robertson organized the events into fourteen categories, distinguishing acts by and against the police, assaults from injuries for which no one was directly responsible, and looted stores from those reported only with broken windows.

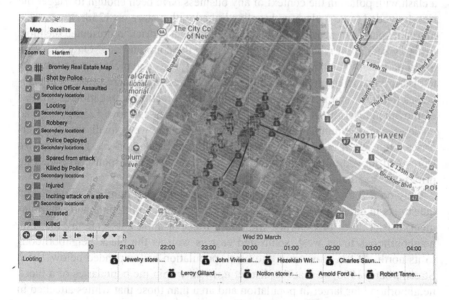

Figure 8.2 Year of the Riot is a deep map of the 1935 Harlem Riot.

Taking into account those filters, more is evident on the map than the attacks on white property and clashes with police that dominate both the initial accounts of the riot and the subsequent historiography.[24] There were at least 26 attacks by Blacks on whites. White store owners, white men and women on the street, newspaper reporters and photographers, and passengers in vehicles traveling through Harlem, all suffered injuries (allegedly at the hands of black assailants). Those assaults are missing from other accounts because they left few traces in the historical record. Attacks on whites occurred throughout the duration of the riot.

However, that violence was more geographically contained than in race riots in the north earlier in the twentieth century: other than two storekeepers attacked in their stores in Harlem's north, most attacks occurred around 125th Street, with a small number further south, around the stores on 116th Street. Moreover, there is very little evidence of the attacks in Harlem or elsewhere by whites, other than police, on black residents that characterized those riots. So, while the riot involved attacks on white commercial property not previously part of public disorder involving African Americans, that is not the full story of what happened on March 19 and 20.

The map provides a basis for shaping new narratives of the riot that pay greater attention to the particularity of the events and to their spatial dimensions. It took a very specific confluence of circumstances and places needed to trigger a novel form of public disorder in Harlem. Not any report or rumor of police violence would have been enough; there were many others in the weeks and months and years preceding March 19, and the violence they provoked was focused on police. Such clashes had never extended to attacks against (white) businesses. Nor would a clash with police in the context of any business have been enough to trigger the riot. It needed to be a clash with police linked with a store on 125th Street. As a journalist put it, for nearly a decade the 125th Street district "represented the most irritating section of white business interests in the community."[25] Only on 125th Street had stores been the target of prolonged boycotts and pickets and continued to discriminate in employment. Only on 125th Street were there venues and businesses that had, and in some cases continued to, practice segregation as well as employment discrimination. Most of the white businesses above 125th Street had made greater accommodations to their customers, and were small, family enterprises, with few if any jobs available for white or black workers. Other uptown stores had moved more quickly to hire black workers in response to picketing prior to the campaigns on 125th Street.

Only on and around 125th Street were there significant numbers of whites on the streets in addition to in businesses. It was the major shopping and entertainment district north of Central Park, and a transportation hub, catering to white neighborhoods to its west, south and east as well as the black neighborhoods to its north. At the same time, the black population had expanded below 125th Street around 1930, so that the district now sat within the boundaries of a black neighborhood far larger in population and area than those that whites attacked in race riots earlier in the century. In addition, thanks to the geography of Manhattan, black Harlem was not surrounded by white neighborhoods as were other black enclaves in the city. Consequently, racial violence on 125th Street would not draw whites from other areas of the city in the way clashes in smaller neighborhoods had. A clash on 125th Street would more quickly involve white members of the CP than one elsewhere in Harlem, as they had a local headquarters just north of 125th Street on Lenox Ave, and the Young Liberators had their headquarters just south of 125th Street on Lenox Ave.

The map also provides a basis for a narrative of the riot that extends beyond the events at Kress and on 125th Street. Events spread in a complex pattern, north, further up Lenox Ave than the other avenues, and south into Harlem's Puerto Rican neighborhoods, in several waves of activity that produced various forms of violence. While the aerial view offered by a map reveals the spread of the riot, "gazing down from a great height makes it hard to see chaos and confusion," as Vincent Brown notes in regard to his map of the slave revolt in Jamaica.[26] In the case of the riot, what we can't see are the crowds that filled the streets for much of the night, and often literally surrounded the events that appear on the map. A point on the map generally represents a moment when groups emerged from the crowds to attack individuals, buildings, or vehicles. However, crowds were present on the streets for more than the moments captured on the map.

As the riot spread over Harlem, events followed a pattern unlike any race riot that preceded it. First crowds gathered to protest, leading to clashes with police; then windows were broken; and finally, sometime later, looting broke out. The map has limits as evidence of the riot's chronology. Only 46% (84/184) of the events on the map appear on the timeline; for the majority is there no information on when they took place. That is particularly the case for stores with broken windows; only 3 of 57 appear in the timeline. Nonetheless, the map does fit with the pattern reported in other sources.

Enough time elapsed between the shifts in behavior that they need to be discontinuous, and each phase of disorder considered in its own right. The progression from one to the next was not inevitable—at least, not in 1935. The period before 10:00 p.m. saw a protest that expanded to clashes with police when their efforts disperse crowds escalated the violence, leading to attacks on Kress' spreading to the other large stores that had been the targets of the boycott movement. A second phase of violence resulted after 10pm in part from police pushing crowds away from 125th Street, on to the avenues that ran north/south through Harlem. Groups of black assailants attacked whites they encountered on the street and in passing vehicles, but they encountered relatively few in those areas of Harlem. In this context, businesses provided an alternative outlet for racial antagonism. That some black-owned stores were caught up in this violence is not at odds with that interpretation (Figure 8.3). As crowds moved through Harlem, not everyone would have been aware of which stores were black owned. When black storeowners put up signs identifying themselves, crowds generally avoided them. Although the shift to breaking windows literally opened the way for looting, that looting did not immediately occur further suggests that attacks on stores were initially an attack on whites.

The time lag between attacks on stores and looting that began after midnight suggests that new groups joined the crowds. Given that the riot occurred in the midst of the Depression, which had hit Harlem residents particularly hard, it is unsurprising that some saw an opportunity to alleviate their economic needs by looting stores that had been targets of racial violence. That the looting appears to have been concentrated on Lenox Avenue fits such an interpretation. The street was in some ways the least likely of Harlem's avenues to be the main target of looting, as it had long been home to lower grade stores than on 7th Avenue, but the blocks to the east were Harlem's poorest and most overcrowded. If much of the looting was the work of the hungry, it did not supplant racial antagonism. As some

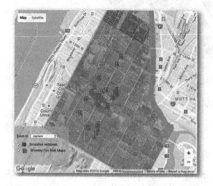 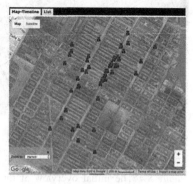

Figure 8.3 Year of the Riot showing broken windows (left) and looted locations (right).

groups looted stores, others continued to attack whites, in a context of intensifying violence fed by police beginning to shoot at looters.

Harlem's riot was thus not a total break with the past. To the extent that they could, Harlem's residents attacked whites. A spatial perspective highlights that the new forms of racial violence that appeared for the first time in Harlem resulted at least as much from the targets available as a change in motives or the nature of racial antagonism. There were fewer whites to attack in the new, larger black neighborhood, and few whites responded to racial violence by venturing into black neighborhoods (and were less motivated to do so when provocations did not involve clashes between black and white populations, but between black residents and police and storeowners). Rather than a break with the past, the 1935 riot involved the layered violence that Dominic Capeci and Martha Wilkerson have argued characterized the later riot in Detroit in 1943: a transitional moment that "piled distinct layers of violence atop one another" to encompass both violence against individual whites, and against white authority and property.

The narrative map first unpacks 125th Street as a context, and then divides the events into four chronological phases, each visible when the timeline is positioned within the appropriate timespan: 2:30–10:00 p.m.; 10:00 p.m. to midnight; midnight to 2:30 a.m.; and 2:30 a.m. to 5:30 a.m. (Figure 8.4).[27] That the exhibit is effectively a series of maps suggests that Neatline doesn't handle complexity any better than Story Map platforms. However, linking those maps and waypoints containing text and images to Neatline's timeline allows for a more dynamic and engaged pathway through those maps. The timeline slider provides the means of navigating the exhibit—dragging it changes not only the points visible on the map but also the waypoints visible in the right-hand menu.[28]

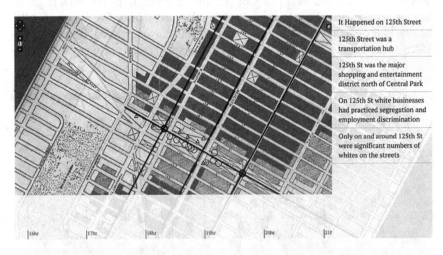

Figure 8.4 The initial overview at the beginning of the spatial narrative of the 1935 Harlem riot shows the narrative structure through Neatline's waypoints (right) alongside the map of events in their spatial context.

   Those waypoints are not tied to particular events; they explore the broader patterns (Figure 8.5). And rather than each being tied to an individual point on the timeline, waypoints elaborating related arguments are grouped together. The window in which each waypoint displays only provides space for a few sentences of text, so grouping them allows scope for more complex ideas to be developed. By the same token, breaking those ideas into smaller sections makes them more accessible, with each section title operating as a marker and a summary (when compared to a larger block of scrolling text). Groupings of waypoints also provide some flexibility in how a narrative is read—you can roll over each waypoint in a group, and explore them out of sequence, with those options limited to reduce the chance that you could lose your way in the argument. And of course, opening the window on the map keeps attention on the map, and brings an element of the idea of a pathway through the map.

   Beyond the text, the waypoints work to provide a pathway through the map in two other ways. Each waypoint can be associated with a zoom level centered on a specific location. Clicking on a series of waypoints can thus move you around a map. Annotations of the map can also be attached to each waypoint. Roberston has tried to use polygons and lines to direct attention to the analysis of space in the narrative: to movement, direction, proximity, connection, and patterns. Used in that way, annotations shift some of argument into a visual form. Combined, zooming and centering and annotations offer guidance on how to read the map, while allowing the map to retain some complexity by keeping all the points from a timespan visible—rather than having to disaggregate event types into layers,

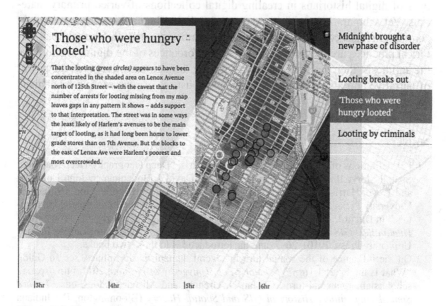

Figure 8.5 The spatial narrative of the riot in Neatline zooms in and displays the relevant items from the deep map, accompanied by narrative text.

as is necessary to highlight patterns in the map from Year of the Riot. Having the visibility of the annotations tied to the timeline also ensures the map is not cluttered and made unintelligible.[29] The narrative that unfolds in the waypoints is essentially independent of the information about the individual points representing events on the map, even though conceptually it depends on it.

The Neatline software is not entirely satisfactory as a platform for spatial narrative, but its flaws are outweighed by the affordances it offers for creating spatial narratives not available in other platforms, with recent NEH funding for its further development promising further refinement and development of those functionalities.[30] Because it is built on the Omeka platform, it can create deep maps with a multiplicity of kinds of sources and arrange them spatially as points on the map. But even more important is its ability to select and arrange the materials into a narrative on a map. Besides the inherent selection and arrangement which goes into adding items to the Omeka site, Neatline requires users to take materials from the Omeka site, select which ones will go into the map, and spatially arrange them as points or waypoints. And it also takes the form of augmenting them with waypoints that frame points or sections of the map and arranging them into the narrative which the user can use to navigate. Of course, Neatline has the conception of having multiple maps or narratives. So even within the universe of (already selected, already arranged) materials within the Omeka repository, the Neatline maps allows for different selections and arrangements of those materials. It is possible, then, to allow a multiplicity of narratives through deep map.

The idea of a deep map is to use the affordances of new media to escape the limitations of printed prose or mapmaking. In this regard, it is similar to the practices of digital historians in creating digital collections of varied primary materials. Yet scholars in spatial history and the digital humanities are increasingly coming to regard collection as a useful but insufficient step. If the traditional historical practices are the thesis, and the new affordances of the digital medium are the antithesis, then the synthesis of navigating through narrative weds the power of narrative and argumentation to the capaciousness of representing humanistic sources on a deep map.

Notes

1 For definitions of deep mapping see Trevor Harris, "Deep Geography-Deep Mapping: Spatial Storytelling and a Sense of Place," in David J. Bodenhamer, John Corrigan, and Trevor Harris, eds., *Deep Maps and Spatial Narratives* (Bloomington, IN: Indiana University Press, 2015), 28; David J. Bodenhamer, "The Potential of Spatial Humanities," in David J. Bodenhamer, John Corrigan, and Trevor M. Harris, eds., *The Spatial Humanities: GIS and the Future of Humanities Scholarship* (Bloomington, IN: Indiana University Press, 2010), 26–7; and the introductions to those two books.

2 On the influence of the spatial turn in several humanistic disciplines, see Jo Guldi, "What Is the Spatial Turn?" *Scholar's Lab, University of Virginia*, 2011. http://spatial. scholarslab.org/spatial-turn/. Cf. Ian N. Gregory and Alistair Geddes, eds., *Toward Spatial Humanities: Historical GIS and Spatial History* (Bloomington, IN: Indiana University Press, 2014). On the ways in which historians have used maps and spatial analysis, and on the distinction between the spatial humanities and spatial history, see

Stephen Robertson, "The Differences between Digital Humanities and Digital History," in Matt Gold and Lauren Klein, eds., *Debates in the Digital Humanities 2016* (Minneapolis: University of Minnesota Press, 2016). https://dhdebates.gc.cuny.edu/read/untitled/section/ed4a1145-7044-42e9-a898-5ff8691b6628.

3 Robert Darnton, "The New Age of the Book," *New York Review of Books* (March 18, 1999). www.nybooks.com/articles/1999/03/18/the-new-age-of-the-book/.

4 Daniel J. Cohen and Roy Rosenzweig, *Digital History: A Guide to Gathering, Preserving, and Presenting the Past on the Web* (Philadelphia: University of Pennsylvania Press, 2015). http://chnm.gmu.edu/digitalhistory/.

5 Edward Ayers and William G. Thomas, III, *The Valley of the Shadow: Two Communities in the American Civil War*, University of Virginia, 1993–2007. http://valley.lib.virginia.edu/; William G. Thomas, III and Edward Ayers, "The Differences Slavery Made: A Close Analysis of Two American Communities," *American Historical Review* 108, no. 5 (2003): 1299–307. See the online version of the article: www2.vcdh.virginia.edu/AHR/.

6 Anne Sarah Rubin, et al., *Sherman's March and America: Mapping Memory*, University of Maryland Baltimore County, 2010–2019. http://shermansmarch.org.

7 For a recent taxonomy of digital history projects and their relationship to argumentation within the field, see Arguing with Digital History Working Group, "Digital History and Argument," white paper, Roy Rosenzweig Center for History and New Media (November 13, 2017). https://rrchnm.org/argument-white-paper/.

8 Edward Ayers, "Does Digital Scholarship Have a Future?," *Educause Review* (August 5, 2013). http://er.educause.edu/articles/2013/8/does-digital-scholarship-have-a-future; Cameron Blevins, "Digital History's Perpetual Future Tense," in Matt Gold and Lauren Klein, eds., *Debates in the Digital Humanities 2016* (Minneapolis: University of Minnesota Press, 2016). https://dhdebates.gc.cuny.edu/read/untitled/section/4555da10-0561-42c1-9e34-112f0695f523; William G. Thomas, III, "The Promise of the Digital Humanities and the Contested Nature of Digital Scholarship," in Susan Schreibman, Ray Siemens, and John Unsworth, eds., *A New Companion to Digital Humanities* (Chichester: Wiley Blackwell, 2016), 603–17.

9 A *New York Times* graphic designer has noted that only about 10% to 15% of visitors to infographics on their site interact with them, even superficially. Meg Miller, "The Problem With Interactive Graphics," *Fast Co. Design* (March 17, 2017). www.fastcompany.com/3069008/the-problem-with-interactive-graphics?. Vincent Brown reports that during a spike of visitors to his interactive spatial narrative *Slave Revolt in Jamaica*, "nearly 90 percent of those visitors did not click past the first page." Vincent Brown, "Narrative Interface for New Media History: Slave Revolt in Jamaica, 1760–1761," *American Historical Review* 121, no. 1 (2016): 176. We do not cite these figures to discredit the idea of interactive graphics or spatial narrative altogether, since of course the 10% of visitors that do engage with these projects often do so at great length and in great detail. But these figures do imply that users are often extremely reluctant to engage with even sophisticated interactive works of scholarship.

10 Harris, "Deep Mapping-Deep Geography," 28.

11 David Bodenhamer, "Narrating Space and Place," in David Bodenhamer, John Corrigan, and Trevor Harris, eds., *Deep Maps and Spatial Narratives* (Bloomington, IN: Indiana University Press, 2015), 17.

12 Bodenhamer, "Narrating Space and Place," 10. Cf. Stephen Robertson, "Putting Harlem on the Map," in Jack Dougherty and Kristen Nawrotzki, eds., *Writing History in the Digital Age* (University of Michigan Press, 2013), 186–97. http://quod.lib.umich.edu/d/dh/12230987.0001.001/1:8/–writing-history-in-the-digital-age?g=dculture;rgn=div1;view=fulltext;xc=1#8.2.

13 On the distinction between *space* and *place*, see Tim Cresswell, *Place: A Short Introduction*. Short Introductions to Geography (Malden, MA: Blackwell, 2004).

14 Bodenhamer, "Narrating Space and Place," 11.

15 Susan Schulten points out that the development of map literacy was a historical achievement with roots in the nineteenth-century development of thematic mapping. Susan Schulten, *Mapping the Nation: History and Cartography in Nineteenth-Century America* (Chicago, IL: The University of Chicago Press, 2012).

16 Much of this chapter was presented as Stephen Robertson, "Toward a Spatial Narrative of the 1935 Harlem Riot: Mapping and Storytelling after the Geospatial Turn," New Approaches, Opportunities, and Epistemological Implications of Mapping History Digitally: An International Workshop and Conference, German Historical Institute, Washington, DC, October 20, 2016.

17 Emily Withrow and Zach Wise, et al., *StoryMapJS*, Knight Lab, Northwestern University. https://storymap.knightlab.com/.

18 ESRI, "Telling Stories with Maps," white paper (February 2012). http://storymaps. esri.com/downloads/Telling%20Stories%20with%20Maps.pdf. In 2019 ESRI released a beta form of an updated platform, ArcGIS StoryMap, which uses blocks rather than templates, but does not appear to change the character of the spatial narratives that can be produced: ESRI, "ArcGIS StoryMaps." www.esri.com/en-us/arcgis/products/arcgis-storymaps/overview.

19 Gregory P. Downs and Scott Nesbit, *Mapping Occupation: Force, Freedom, and the Army in Reconstruction*, 2015. http://mappingoccupation.org. For quotations, see "about" page.

20 Scholar's Lab, *Neatline: Plot Your Course in Space and Time*, University of Virginia, 2009–2017. http://neatline.org/.

21 Stephen Robertson, Shane White, and Stephen Garton, *Year of the Riot: Harlem, 1935*, 2011–2019. http://digitalharlem1935.org; Stephen Robertson, Shane White, Stephen Garton, and Graham White, *Digital Harlem: Everyday Life, 1915–1930*. http://digital harlem.org/. For the prototype spatial narrative of the riot, see http://1935harlemriot. org/narrative/neatline/fullscreen/march-19-20-1935.

22 See "Digital History and Argument," 1–2.

23 We use the word *prototype* deliberately, since working in Neatline is not part of the larger *The Year of the Riot* project. It is a sideline inspired by the opportunity offered by Robertson's talk at the German Historical Institute (cited previously) and Mullen's participation in the deep mapping workshop at West Virginia University.

24 Mayor's Commission on Conditions in Harlem, *The Negro in Harlem: A Report on Social and Economic Conditions Responsible for the Outbreak of March 19, 1935* (New York, n.p., 1935); Claude McKay, *Harlem: Negro Metropolis* (New York: E. P. Dutton, 1940); Roi Ottley and William J. Weatherby, *The Negro in New York: An Informal Social History* (New York: Oceana Publications, 1967), 275–80; Mark D. Naison, *Communists in Harlem during the Depression* (Urbana: University of Illinois Press, 2005), 140–7; Cheryl Greenberg, *Or Does It Explode: Black Harlem in the Great Depression* (Oxford: Oxford University Press, 1991), 5; Cheryl Greenberg, "The Politics of Disorder: Reexamining Harlem's Riots of 1935 and 1943," *Journal of Urban History* 18, no. 4 (1992): 395; Lindsey Lupo, *Flak-Catchers: One Hundred Years of Riot Commission Politics in America* (Lanham, MD: Lexington Books, 2011); Lorrin Thomas, *Puerto Rican Citizen: History and Political Identity in Twentieth-Century New York City* (Chicago, IL: The University of Chicago Press, 2014), 75–83; Marilynn S. Johnson, *Street Justice: A History of Police Violence in New York City* (Boston: Beacon, 2004), 186–9; Allen D. Grimshaw, "Lawlessness and Violence in America and Their Special Manifestations in Changing Negro-White Relationships," *Journal of Negro History* 44, no. 1 (1959): 52–72; Morris Janowitz, "Patterns of Collective Racial Violence," in Hugh Graham Davis and Ted Robert Gurr, eds., *The History of Violence in America* (New York: Praeger, 1969).

25 *Atlanta World*, March 27, 1935, 2.

26 Vincent Brown, *Slave Revolt in Jamaica, 1760–1761: A Cartographic Narrative*, 2012. http://revolt.axismaps.com. For the quotations, see under heading "Uncertainty."

27 The map uses more opaque borderless points to represent events that can't be placed on the timeline; they appear after 10 p.m. and remain on through the remaining phases of the riot, the timespan in which they would have occurred.

28 The timeline functionality relies on an old, buggy, and limited SIMILE widget that couldn't display the number of records related to the riot, and which had to be broken to prevent it from jumping forward in a way that separated records from the point to which they refer on the map.

29 However, Neatline does not make it possible to integrate points and waypoints—only one record can be associated with a waypoint. So, while one can associate many points and polygons with a waypoint, each cannot have its own record, so multiple events cannot be linked to a waypoint without losing information about each. (Waypoints are not an overlay on other records, but a record on par with them.)

30 Neatline has recently been awarded NEH funding to update it and expand its functionality. Eric Rochester, "Neatline Implementation Grant," *Blog Post*, August 12, 2016. http://scholarslab.org/announcements/neatline-implementation-grant/.

9 Cultural heritage institutions and deep maps

Mia Ridge

Introduction

Does deep mapping fall victim to the dearth of digital data? The question was posed at a workshop on "Framing the methodological foundations for deep mapping and spatial story telling" at West Virginia University in Morgantown in May 2016. Museums, libraries, and archives are rich in collections of historical maps, images and texts that could help deepen a "deep map" and have been busily digitizing them over the past 20 years. But the answer—for now—may be "yes," as despite the apparent deluge of digitized data, much of it is not (yet) ready for inclusion in deep maps.

This chapter addresses some of the reasons why digitized collections suitable for use in deep maps are not yet as freely available as one might expect, given this investment in digitization. It explores the impact of organizational contexts in cultural heritage institutions (CHIs)[1] on the development of deep maps. Through a consideration of their collecting, cataloguing and digitization histories, resourcing and technical constraints, and missions or goals, it will help users of deep maps understand the processes that shaped the publicly available outcomes of digitization projects. It reflects on the ways in which methodological considerations and practical decisions about representing data intersect and accumulate as spatial collections are digitized and published. Illustrated with examples from the British Library, this chapter is grounded in the author's experience with digital projects in a range of museums, libraries, and archives, particularly in the Australian and British cultural heritage sectors.[2] While it is a snapshot of a particular moment in time, with details liable to change, the impact of the twin constraints of limited resources and the growing scale of cultural heritage collections are likely to remain consistent over time. Understanding the organizational context provides insight into apparent barriers to the creation of or contribution to deep maps. This greater understanding of the issues shaping digitized collections may help remove some barriers to the development of deep maps.

The advantages of deep maps for cultural heritage institutions

Deep maps, most straightforwardly defined as "finely detailed, multimedia depiction[s] of a place and the people, animals, and objects that exist within it,"[3]

DOI: 10.4324/9780367743840-9

could be ideal environments for accessing many cultural heritage collections. As interfaces designed to encourage the exploration of specific places, they can display texts, images, and multimedia to advantage in an environment tailored for rich resources. As engaging experiences that encourage deeper curiosity about the lived experiences of other times and places, deep maps are a good match for the mission of most CHIs. In contrast to the experience of browsing thumbnail images or lists of titles, maps and place-based interfaces seem to provide an immediate sense of relevance. As innovative platforms designed to encompass the uncertainty, fuzziness, contingency, and multivocality contained within cultural heritage collections, deep maps could provide useful working models for other cultural heritage collection interfaces. The emphasis on place and space in deep maps is ideal for reuniting historic collections that have been divided between different institutions, departments, or catalogues over time. For example, the collections of Hans Sloane (1660–1753) formed the basis of the British Museum's library (which subsequently became a foundational collection for the British Library), but some objects were later moved to the Natural History Museum. Moreover, some of Sloane's books were sold to other libraries during a series of duplicate sales in the late eighteenth and early nineteenth centuries, further dispersing his collection.[4] A deep map platform could provide an experience that reassembles those collections and helps researchers understand them as originally constituted. By providing more information about the temporal, spatial, and cultural context in which items were collected, deep maps could help researchers understand the composition of collections. They could perhaps even make visible the shadow of items that escaped the collector or did not survive the passage of time. While it may challenge some, the reflexivity emphasized in discussions of deep maps by Bodenhamer, Corrigan, and Harris is not antithetical to institutions aware both that their historical collections are in some ways happy accidents of survival, and that their cataloguing practices reflect the views of contemporary society as much as they do the item itself. In short, deep maps could provide innovative and engaging interfaces for spatially indexed cultural heritage collections.

So why are we yet to see deep maps supported by cultural heritage institutions? One reason may be that some aspects of deep maps can seem frustratingly slippery to the practically minded digitization officer or curator. Cultural heritage organizations tend to be delivery-focused, designing products—whether exhibitions or online catalogues—within tight timelines for the widest possible audiences. CHI projects are delivered by multi-disciplinary teams, which may include project managers, software designers and developers, user experience researchers, exhibition designers, cataloguers, photographers, learning, and marketing staff as well curators. CHI staff are thoughtful about the products they create, but do not usually have time to theorize about emerging technologies or invest in as-yet-unproven platforms. Deep maps are also posited as "intentionally subversive, imprecise, complex, reflexive, sumptuous, and resplendently untidy"[5] while including "the discursive and ideological dimensions of place"[6]—qualities less likely to fit into a project management plan or to be achievable without significant resources devoted to bespoke design and programming. The resources required to create such a deep map may also draw on the collections of more than one institution and may therefore be seen as not within the remit of any one institution. Yet some may question whether deep maps are truly deep maps if they do not involve some form of negotiation over what is represented and how different self-contained collections are included. The unresolved tension between those definitions of deep maps that focus on their ability to surface information and historical records, and those that focus on deep maps as sites for new experiences of historical materials and past times could mean that some deep-ish maps created by CHIs may not be recognized as such. CHIs may see the benefits of "visual, time-based, and structurally open" and "genuinely multi-media and multilayered"[7] deep maps, but on a number of levels they may struggle with the idea of interfaces that expect negotiation between "insiders and outsiders, experts and contributors, over what is represented and how."[8]

Audiences, mission, and goals

Understanding typical audiences, missions, and goals that underlie decision-making and prioritization can be useful when dealing with cultural heritage institutions. Nick Poole's 2012 analysis of mission statements from a range of UK museums found that their first duty is to "people, society, audiences [and] users," and then to "collections."[9] Collections underpin the work of cultural heritage organizations in inspiring, educating, and entertaining their various publics. Their value lies in their use and accessibility, but this must be carefully balanced with their long-term preservation. For example, the British Library's mission is to "make our intellectual heritage accessible to everyone, for research, inspiration and enjoyment"; this mission then helps prioritize the many requests for collaboration or digitization.[10] CHI mission and goals are translated into programs of activity with implicit or explicit metrics for success. For some institutions, success might be measured in the numbers of physical visitors to reading rooms and exhibitions, online resources accessed, images used by documentary makers or artists, or publications citing collections. Others measure income generated, or the impact and amount of the reuse of digitized images, texts, and multimedia in creative, commercial, or educational projects. Understanding the metrics used by a specific CHI may help would-be collaborators phrase any requests in terms that make the benefits of a potential project clear for both parties.

Digitized items, when licensed for scholarly and creative reuse, can help organizations fulfill their mission. Digitization can make collections accessible to local, national, and international audiences, 24 hours a day. Digitization has other benefits, helping protect fragile items (such as historical newspapers) by reducing the amount of handling they receive. While it was feared that digitization would reduce demand for in-person access and experiences, in practice digitization enhances the reach of cultural heritage collections. On a practical level, digitized collections are available almost instantly, regardless of the reader's physical location, and digital access has overtaken in-person access to collection items. For example, in one year at the British Library, 400,000 Reading Room sessions led to over 1.65 million items being accessed, but just under 3.7 million items were consulted on the British Library's websites.[11] Digitized items can appear in multiple locations simultaneously, and as a general rule, the more open the license, the greater the number of academic and public projects in which it may be used. Digitized items published online with an open license can enable collaboration without paperwork, a huge benefit considering the logistical and legal issues that may otherwise arise. For example, a scholar can include open access collections in a specialized interface alongside items from other institutions without the need to negotiate an inter-institutional contract. The same open access items could be ingested into Wikimedia Commons for inclusion in Wikipedia articles, reaching a far wider audience than on any institution's own site. For example, in June 2017 over 42 million pages with images from the British Library were accessed across the entire Wikipedia site. In the same month, Gerardus Mercator's 1554 map of Europe was viewed 112,145 times on Spanish-language Wikipedia,[12] far exceeding its reach on the British Library website.

Shaping digital collections: collecting, cataloguing, and digitization

Considering the history of collecting, cataloguing, and digitizing collections at institutions like the British Library can help users of those collections understand the impact of past decisions on the discoverability and reusability of items today.

Collecting

The collections of the British Library are vast, with an estimated 180–200 million items ranging from Chinese oracle bones over 3,000 years old to websites crawled for the UK Web Archive today. The British Library holds about 4.5 million maps from the last 2,000 years, including manuscript and printed maps, and more recently, born-digital maps. The richness contained within individual items offers tantalizing opportunities—manuscript and printed pages, sound and vision archives, and extensive collections of images are potentially dense with references to real and imagined spaces. However, the number of collection items creates almost unique challenges of scale for almost any aspect of digitization, cataloguing, access, and preservation.

Map collections may be grouped by source organization—for example, print and digital maps from the Ordnance Survey, items from the India Office Records, the Ministry of Defence, fire insurance maps and shopping center plans. Others are grouped by collector, including the Crace collection of maps of London, and the foundational collections of Hans Sloane and the Cotton and Harley families. Other significant collections include King George III's maps and views (about 50,000 manuscript and printed maps 1540–1824, plus architectural drawings and views). Other collections are the outcome of legal processes, including legal deposit, the requirement to deposit a copy of new British publications with the British Library. Maps contained in books acquired via legal deposit may be catalogued differently than printed "single sheet" maps; others may be collected as drawings or views rather than as maps. Collecting histories can therefore, also affect how spatial content is catalogued and subsequently, when and where it becomes available. Finally, different collections may have different copyright and license terms, which then determine whether the digitized version can be made available for reuse.

To an extent, the scope of a single collection can be described by listing the places it depicts in different formats (maps, views, etc.), but it is less straightforward to describe the ways in which it has been shaped by the interests of collectors, or the physical factors that led to the items within it being created, preserved, and collected. For example, the Crace Collection of Maps of London was shaped by the interests of Frederick Crace (1779–1859). Crace collected different editions of the same map, and commissioned both topographical images of London buildings and copies of maps that he was unable to buy, providing a version of maps that do not survive in the original.[13] His interests allow historians to compare proposed developments with actual ones, and to trace changes in specific suburbs over time. The scope and shape of a collection influences potential uses and users. Thus, understanding the provenance of a single map and information about the collectors who valued it can be important for understanding the places and spaces it represents, and consequently for displaying it appropriately in a deep map.

Cataloguing and discoverability

Library catalogues were designed to help readers find relevant items and locate them on the shelf or order them from closed stacks. The metadata recorded in catalogues is crucial for discoverability, the ability to find a specific item or one of many items with information that can answer the users' question. Catalogue data are usually structured with indexed fields for creators (authors, editors, illustrators, etc.), publication details (place, date, etc.), and subjects, and other fields less likely to be used in search, such as dimensions and bindings. The systems that store this metadata were originally designed for analogue material and are based on analogue systems. The card catalogue may have disappeared from reading rooms, but the space limitations of an index card are still present in cataloguing conventions.

Catalogue records are often palimpsests, containing the traces of decades of past practices. The scale of cultural heritage collections and the cost of manually creating metadata mean that many legacy records in CHIs are sparsely populated, limited to information readily available on the title pages of items. The subject headings applied to items were designed for the physical library environment, where assistance from librarians was available, rather than the online environment where users are unable to ask for help as easily.[14] More importantly for deep maps, traditional catalogue records cannot take full advantage of the affordances of digitized items. For example, images from the pages of a digitized book can take on a life of their own, gathering annotations in the form of crowdsourced tags, scholarly marginalia, citations, or links to pages in other works. These annotations could improve the discoverability of individual items, but traditional catalogues cannot record information about specific pages, let alone regions of pages.

Many library systems are based on data from index cards that has been converted into electronic formats, and subsequently undergone several system migrations. Cataloguing collections at scale, with reduced resources,[15] is challenging. Cumulative decisions about how to manage limited resources and how to map existing fields into newer standards affect both the quality of search results and the reusability of catalogue data in external applications. Users trained by Google's "automagical" ability to understand what they mean by a search term and give them the result they seek can be baffled by traditional CHI catalogues or finding aids. To further complicate matters, libraries and archives catalogue and treat prints and drawings differently from museums or galleries.[16] Reusing metadata for computational processes beyond simple searches can be difficult. Most metadata fields are primarily designed to be found and read by people. Unlike computers, the human reader is not flummoxed by small inconsistencies, uncertain or fuzzy values or by multiple values within a single field. Consequently, legacy metadata may not be consistently structured in ways that support computational processes. Consistent data are needed to drive browse or search interfaces. Accurate, consistent metadata is vital for deep maps—without location data, items cannot be correctly placed on a map.

The British Library's catalogues reflect the age and variety of its founding collections, with different bibliographic approaches used for different collections still reflected in the metadata available today.[17] The British Library also has over 20 specialized catalogues that help people find collections relevant to their purpose. These catalogues contained extended information on specific attributes relevant to, for example, illuminated manuscripts, georeferenced maps, Victorian ephemera, or book bindings. However, as these databases are separate from the main databases, these fields are not available for search in the core catalogues alongside common fields like title, creator, or place of publication. Decisions made about which system something is catalogued in can have a dramatic impact on the usability of collections in deep maps.

Digitization projects usually require extra staff time to modernize and enhance catalogue records to meet current standards, and digitization projects for deep maps may require even more extra resources. The standards for cataloguing cartographic items include attributes not recorded for non-cartographic items, including coordinates, scale, projection, presentation technique, representation of relief (e.g., contours, shading, or hachures), the recording technique (e.g., for remote sensing images), and other characteristics.[18] Unlike books, single-sheet maps and views may not have an obvious title or author, and finding information on the publisher, date, and place represented may require further investigation. Maps may mention surveyors, cartographers, editors, or engravers; each of which should be noted. This highly structured information is usually amenable to computational processing, particularly when gazetteers and standard vocabularies are used to disambiguate people and places from others with similar or identical names. However, recording these additional or more complex attributes increases the time required to catalogue each item, thereby increasing the costs for geospatial digitization projects.

Digitization

The simple term "digitization" hides the complexity of the processes required to digitize an item or collection. The data points underlying the Digitization Cost Calculator (http://dashboard.diglib.org/data) maintained by the Digital Library Federation Assessment Interest Group's working group on Cost Assessment gives a sense of the many tasks that must take place before, during, and after digitization. They include rights clearance and licensing, removing fastenings or unbinding items, quality control, post-capture processing and metadata creation. The composition of the collection to be digitized is a factor in the cost, as collections with items of mixed sizes and format take longer to set up than more homogenous material. The steps described by staff working on the British Library's Kings Topographical Collection cataloguing and digitization "K.Top" project, which aims to provide free online access to 30–40,000 maps and views collected for King George III, are typical. First, the size and content of the collection are assessed; conservation treatment is organized for items that need it; and items are photographed or scanned (depending on their size, condition, format, the project budget, and timelines). Catalogue entries are created or updated to include full transcriptions of titles and linked "name authorities" for those involved in its creation (including artists, printmakers, and publishers) or collection (previous owners and other provenance information). Finally, the digitized items are ingested into a digital library system for preservation and publication online.[19] Digitizing a collection of 40,000 maps and views takes time, with perhaps 20 items a day catalogued, and 60 items scanned a day. In the case of K.Top, a related project, "Transforming Topography," was funded to host events, support researchers and write articles for a new website called "Picturing Places" (www.bl.uk/picturing-places). This work was only possible because of the generous financial support of several donors.

The British Library has been digitizing collections for 20 years, but it is estimated that at most 4% of the collections are digitized or born-digital. In most CHIs, digitization is funded through specific projects supported by government grants, academic or commercial partnerships, or philanthropic donations. Accordingly, funds for digitization may be affected (reduced, usually) by economic and political changes. Meanwhile, organizations with a remit to collect contemporary material must keep acquiring new items that then join the digitization backlog. Improvements in digitization standards have led to changing user expectations, and images of the earliest digitized items (often the "treasures" of a collection) may no longer be considered suitable for modern use and must be re-digitized. For example, at one point, an image of 507 kb with dimensions of 837 px × 1385 px at 72 pixels/inch was considered reasonable for printing out as a reference image but is now only just large enough for an iPhone 6 screen.[20]

Finally, it is important to note the impact of copyright and data protection laws on what can be published online. Copyright and data protection legislation limits access to some nineteenth- and twentieth-century materials, and many are surprised to learn that, strictly speaking, unpublished manuscripts in the United Kingdom—no matter how many centuries old—are copyright until 2039.[21] Hundreds of millions of items are considered "orphan works" as it has not been possible to trace a rights holder who could give permission for their use, complicating the process of sharing them online.[22] Items digitized through commercial partnerships may be held behind pay walls for a certain number of years, limiting their use in deep maps until that license expires.

The usability of digitized resources for deep maps

While there is now a critical mass of digitized data, copyright and technical limitations mean that it is not (yet) easily used in interfaces such as deep maps. Legacy sites from earlier digitization projects are often limited in scope, and the resources they contain are not clearly available for reuse or remixing with other collections. However, ongoing work on technical infrastructure and enhancing records with linked open data should mean that more data are available to creators of potential deep maps.

Scholars of early modern text and images may be happier with the digitized collections available for reuse than scholars of audiovisual material. Copyright clearance is even more complex for sound and video, as information about performers, composers, filmmakers, authors, artworkers, and even the license holders for other copyright material visible or audible in shot must be researched and evaluated.

As outlined previously, specialist digitization projects may involve map cataloguers, but cartographic images in general manuscript or printed heritage collections may not receive the same treatment and may therefore not be as easily found and used by those constructing deep maps. Many maps now available digitally from the British Library may not have previously been mentioned in the catalogue record for the book that contained them. However, in an example that provides a possible model for other projects, the British Library Labs team extracted approximately one million images from digitized nineteenth-century books and made them available for reuse via Flickr Commons (www.flickr.com/photos/britishlibrary).[23] Through a variety of manual and computational processes, volunteers reviewed every image and found over 50,000 maps.[24] These maps are now tagged with page and image-level metadata, suitable for use in deep maps.[25]

Finding collection items for use in a deep map generally relies on item-level metadata or on resource-intensive searches through collections likely to cover relevant places. Metadata systems originally designed to enable orders to reading rooms lack the fine-grained detail that might help place a section of a larger item on a deep map—most metadata can tell you about the volume or box, but not the page. There is a difference in granularity between the catalogue record designed to deliver items to reading rooms or issue desks, and the data gathered about images on specific regions of specific pages on digital platforms. This means that integrating different forms of data is an emerging issue for many libraries that wish to make the most of new sources of data about collections.

The metadata that is available may not be usable in deep maps without further processing, as place names and personal and organizational names may be ambiguous and difficult to match to geospatial coordinates or other sources of information. While the human reader may assume that a reference to London in the title of a book is about the capital of the United Kingdom and therefore roughly equivalent to N 51°30′31″ W 0°07′33″, software is less likely to be able to automatically and authoritatively resolve the string "London" to a specific coordinate, as there are several London's in other countries. Metadata is more easily reusable when it uses standard identifiers linked to external authorities such as the Getty Thesaurus of Geographic Names (now available under an Open Data Commons Attribution License), or other gazetteers specific to a particular subject, place, or period. The Pelagios project provides a model for creating identifiers for places in collections related to the classical and ancient worlds, a model which has been followed for PeriodO, an emerging gazetteer of scholarly definitions of historical, art-historical, and archaeological periods. Work on the World Historical Gazetteer is also promising. Other useful identifiers include VIAF (Virtual International Authority File) for bibliographic data and ISNI (International Standard Name Identifier) for people and organizations involved in making or distributing creative works. These services combine information from a range of CHIs internationally and are designed to support national and regional variations in authorized forms, language, script, and spelling. Records published with linked open data identifiers can be pulled into a deep map by software told which spatial and bibliographic attributes are relevant to that deep map.

The benefits of linked open identifiers are clear but adding identifiers to existing catalogue records is not straightforward. Most cataloguing systems require some technical changes to allow them to record URLs or other identifiers instead of or alongside text fields. The technical work to integrate linked open data identifiers into British Library catalogues is ongoing, but the work of updating catalogue entries will take longer. Some reconciliation work—searching online authorities for likely matches for specific terms or converting standard vocabulary terms to their online equivalent—can be done computationally with tools like OpenRefine, but some work manually searching for and verifying potential and missed matches is inevitable. Meanwhile, emerging machine-learning based computational methods for parsing entire digitized text or images, finding potential place names or other spatial references, and linking them to online gazetteers may reduce the reliance of deep mappers on item-level metadata in future. While a full-text keyword search may find references to Orange County, it may not be clear which state's Orange County is meant, and many other references to "orange" may clutter the results.

Geoparsing, the process of identifying place names then matching them to geospatial coordinates, is particularly relevant for deep maps. However, these emerging text and data mining methods for identifying entities such as specific places, people, organizations, events, or concepts within texts or images cannot (yet) be unproblematically applied to historical items, nor typically those from a non-Anglophone culture. Modern gazetteers such as GeoNames which could be used to generate lists of place names for entity recognition software do not contain the many past place names necessary to represent historical collections. They also cannot easily accommodate imaginary, synecdoche (e.g., "the Somme"), temporary place names, or specialist place names like trench names recorded on documents from the First World War, a problem tools like Pelagios are designed to solve. Digitized versions of historical texts are increasingly available—for example, from the British Library's dataset portal (https://data.bl.uk), the Internet Archive (https:// archive.org/details/texts), the British Library's International Image Interoperability Framework (IIIF) standards-based item viewer[26] and other open access sites. However, the transcribed texts available on these sites are usually automatically created using Optical Character Recognition (OCR) software,[27] which is far more likely to contain errors than manually transcribed text. Gregory et al. provide some insight into the difficulties of turning digitized collections into data points for deep maps as they used parts of the British Library's digitized nineteenth-century newspaper collections in historical geography research projects. They described the significant amounts of work required before exploratory analysis could begin, including attempts to correct OCR errors and to understand their impact on their research process before they could proceed with geoparsing the texts.[28]

OCR technology is constantly improving, aided by developments in machine learning and so-called artificial intelligence. Machine learning is also being applied to train software to transcribe handwritten text in the Transkribus (https://transkri bus.eu) project. The software is "trained" to read a hand by loading it with high-quality manual transcriptions and the related digitized pages. The British Library is experimentally using this technology with records from the India Office Records, and while handwritten text recognition (HTR) is still a developing technology, Transkribus recently transcribed the first page from this collection without any transcription errors. When the accuracy of automatic text transcription improves, it is hard to understate the extent to which it will transform our access to historical collections. Full-text transcription enables text and data mining methods that will provide researchers with computational methods for processing collections in bulk. Full-text transcription also enables keyword searches in digitized items, allowing any obscure word to become a search result. Scholars accustomed to using keyword searches in Google Books should be able to imagine the impact of full-text search in other CHI collections. However, as it also creates challenges for managing the sensible display of far greater numbers of search results, it may be some time before public-facing catalogues integrate this alongside metadata-based searching.

In summary, the usability of digitized resources for deep maps is not yet ideal—deep maps as currently described implicitly rely on either the expensive and resource-intensive manual compilation of text and images about a place, or on computational processing. Improvements in computational methods for recognizing and identifying spatial references should mean that deep maps soon can draw on passages of text or images deep within collection items.

Approaches to spatial infrastructure

With the exception of the National Library of Scotland, whose Map Images pages (http://maps.nls.uk) offer a range of special projects and map viewers, very few cultural institutions in the United Kingdom seem to have the resources to create specialized spatial search and manipulation interfaces for their collections. In the British Library, maps are listed in the main catalogues for its printed heritage, and archives and manuscripts collections; specific place names included in catalogued metadata are discoverable via text-based searches. In the following section, I discuss some projects in which the British Library has collaborated with external bodies to provide more geospatially aware access to collections than the text-based catalogue. Through these projects, users of the British Library's collections have access to an assemblage of spatially focused interfaces; however, the discoverability of these specialist interfaces and datasets from the Library's main sites is not always straightforward. The "Catalogues and Collections" heading in the website's navigation bar does not include every specialist catalogue or database that could be included, nor digitized items available on third-party sites like Flickr or Wikimedia. Some are described as "Online exhibitions" or online galleries, a term that may confuse the user.[29] Work to improve the discoverability of digitized resources across institutions' various microsites, legacy, and partnership projects is an ongoing challenge.

A key project providing access to British Library maps is their implementation of the Georeferencer tool (www.bl.uk/georeferencer), which crowdsources the process of adding control points to historical maps by comparing them to modern maps. These control points then allow the coordinates and scale of the historical image to be recorded.[30] Georeferenced maps are searchable on Old Maps Online (www.oldmapsonline.org), thanks to funding from Jisc, and viewable via a Google Maps-based interface (www.bl.uk/georeferencer/georeferencingmap. html) that shows each historical map as a point on a modern base map. In 2017, this georeferenced data were integrated into the Library's core catalogue.[31] The Georeferencer site is currently focused on the 50,000 images found via the British Library Labs project discussed previously. Volunteers previously georeferenced 2,500 Goad fire insurance plans, Ordnance Surveyor's drawings and the Crace Collection of maps of London from the British Library's collections.[32]

Documents including fourteenth- and fifteenth-century mappae mundi, itineraries, and portolan charts digitized for the Pelagios project are available for download from the British Library's data portal (https://data.bl.uk/pelagios). These and other manuscript and printed maps have been prepared and uploaded under various headings to Wikimedia Commons (https://commons.wikimedia. org/wiki/Category:British_Library). To pick just one example, digitized maps from the War Office related to British East Africa (1890 and 1940; modern-day Kenya, Uganda, and adjacent parts of Tanzania, Burundi, Rwanda, DR Congo, South Sudan, Ethiopia, and Somalia) are available via Wikimedia Commons. The collection includes small sketch maps made by intelligence officers, surveyors' field sheets, and maps annotated in use. Sharing these items via Wikimedia Commons helped the British Library meet the funders' requirement to make them freely available at a time when the Library's own infrastructure to do so was still in development. The Kings Topographical Collection digitization project mentioned earlier will result in additional openly licensed maps and views. The

British Library also has a range of "sound maps," including recordings of world music, regional accents and dialogue, wildlife, and environmental sounds (http:// sounds.bl.uk/sound-maps). Other spatial content that may one day be available for reuse includes the Catalogue of Photographically Illustrated Books and the international Endangered Archives project (http://eap.bl.uk). While the British Library has not had the resources to create a specialized spatial search interface for its collections, the sharing of digitized resources with other sites provides a level of access to resources usable in deep maps.

Conclusion and recommendations

This chapter has discussed several reasons why digitized collections are not yet easily usable in deep maps, and some reasons why the situation may improve as text and data mining technologies improve. It also outlined some potential barriers to cultural heritage institutions investing in deep map technologies, despite the apparent advantages of deep maps for cultural heritage institutions. CHIs tend to be pragmatic in their decision-making, conscious of the fact that they are spending public money to create digital projects that serve the widest possible range of users. While deep maps are still largely theoretical, and have yet to prove their intellectual accessibility to the general public, it is difficult for them to justify the resources that would be required to explore the still-nascent format of deep mapping platforms other than through research funding.

Drawing on previous models for helping cultural heritage institutions understand new platforms and uses for open data, one possible solution is embedding "deep mappers in residence" in CHIs. For example, in 2012–2013, the British Library hosted a Wikipedian-in-residence, Andrew Gray, who organized staff training, "editathons" focused on improving records about specific collections, and who generally encouraged an appreciation of the benefits of open content. Other projects have funded researchers who worked within the British Library to identify and catalogue items for digitization, providing opportunities to discuss and demonstrate their project for Library staff.

Creating and documenting exemplar deep maps through articles, blog posts, and conference presentations is one way to demonstrate the value of deep maps to CHIs. The discussion of audiences, mission, and goals suggests that commissioning and sharing evaluation on the efficacy of deep map interfaces for meeting audience-specific goals and providing immersive access to collections could help justify the resources needed for deep maps. Finally, the technical and design lessons learnt from pilot projects would be useful for both CHIs and academic partners in assessing the scope and cost of future deep maps.

Prototypes are immensely valuable. Small-scale demonstrators could be created using already-digitized collections and existing software libraries, or with more investment and supportive funders, larger scale prototypes could commission new digitization, interface design, and functionality. As the number and variety of easily reusable digitized items grows, it should be possible to create engaging and/ or useful deep maps that inspire CHIs and others to take full advantage of this emerging medium. It is hoped that this chapter provides some insight into how to collaborate with CHIs as providers for collections for deep mapping projects.

Notes

1 Cultural heritage institutions may also collectively be called GLAMs (galleries, libraries, archives, and museums) or "memory institutions."

2 In a nod to the reflexivity of deep maps, I should mention that my own interest in deep maps stems from years working on access to digitized collections for specialists and the public, and from a long-standing fascination with the potential of digital platforms to revitalize the antiquarian notion of "chorography" (the detailed description or delineation of a particular region from many sources). My work with the Digital Scholarship team at the British Library (a cross-disciplinary team supporting the creation and innovative use of British Library's digital collections) has provided many opportunities to reflect on barriers to the use of collections in emerging platforms and other forms of computationally enabled scholarship.

3 David J. Bodenhamer, John Corrigan, and Trevor M. Harris, "Introduction: Deep Maps and the Spatial Humanities," in *International Journal of Humanities and Arts Computing* 7, nos. 1–2 (2013): 174.

4 James Freeman, "Tracing Hans Sloane's Books: A PhD Placement Opportunity." https://britishlibrary.typepad.co.uk/untoldlives/2016/02/tracing-hans-sloanes-books-a-phd-placement-opportunity.html.

5 Trevor M. Harris, John Corrigan, and David J. Bodenhamer, "Conclusion: Engaging Deep Maps," in David J. Bodenhamer, John Corrigan, and Trevor M. Harris, eds., *Deep Maps and Spatial Narratives*. Series on Spatial Humanities (Bloomington, IN: Indiana University Press, 2015), 223–33.

6 Bodenhamer, et al., "Introduction," 174.

7 David J. Bodenhamer, John Corrigan, and Trevor M. Harris, "Introduction," in David J. Bodenhamer, John Corrigan, and Trevor M. Harris, *Deep Maps and Spatial Narratives*, 4.

8 Bodenhamer, et al., "Introduction," 175.

9 Nick Poole, "What Are Museums For?" https://britishlibrary.typepad.co.uk/untoldlives/2016/02/tracing-hans-sloanes-books-a-phd-placement-opportunity.html.

10 The British Library, "Living Knowledge: The British Library 2015–2023." https://librarylearningspace.com/living-knowledge-british-library-2015-2023/.

11 British Library Board, "British Library Annual Report and Accounts 2016/17," 17. www.bl.uk/about-us/governance/annual-reports.

12 Wikimedia, "Page Views on Wikimedia for Category 'British Library'." The Spanish page containing Mercator's map is https://es.wikipedia.org/wiki/Europa.

13 Peter Barber, "Crace Collection of Maps of London: Curator's Introduction." www.bl.uk/onlinegallery/onlineex/crace/.

14 Janet Ashton and Caroline Kent, "New Approaches to Subject Indexing at the British Library." www.researchgate.net/publication/319371151_New_Approaches_to_Subject_Indexing_at_the_British_Library.

15 Ashton and Kent, "New Approaches to Subject Indexing at the British Library."

16 Felicity Myrone, "Prints and Drawings at the British Museum and British Library." www.bl.uk/picturing-places/articles/prints-and-drawings-at-the-british-museum-and-british-library.

17 Ashton and Kent, "New Approaches to Subject Indexing at the British Library."

18 Examples for the Resource Description and Access (RDA) format taken from Yale University Library training material http://web.library.yale.edu/cataloging/cartographic. Further insight into the complexities of cataloguing can be gained from their page on the "Minimal Level Bibliographic Record" for cartographic items http://web.library.yale.edu/cataloging/cartographic/minimal-level-record.

19 The British Library, "Kings Topographical Collection Cataloguing and Digitisation." www.bl.uk/projects/kings-topographical-collection-cataloguing-and-digitisation.

20 The image appears on a little-used section of the BL website so old it still includes a call to share the page (www.bl.uk/onlinegallery/onlineex/maps/uk/004890682.html) on the defunct bookmarking site delicious.

21 Intellectual Property Office, "Consultation on Reducing the Duration of Copyright in Unpublished Works." https://assets.publishing.service.gov.uk/government/uploads/system/uploads/attachment_data/file/368811/consultation-on-unpublished-works.pdf.

22 "Impact Assessment Report on Orphan Works." www.copyrightuser.org/understand/exceptions/orphan-works/.

23 Ben O'Steen, "A Million First Steps." https://britishlibrary.typepad.co.uk/digital-scholarship/2013/12/a-million-first-steps.html.

24 Kimberly Kowal, "Found: More Maps Than We'd Reckoned." The maps can be viewed at www.flickr.com/photos/britishlibrary/sets/72157638735426654/.

25 However, these decontextualized images may raise new questions about the ability of those who encounter them to distinguish between "accurate" and "imagined" scenes of a particular place. Like any image, maps created to illustrate books will leave out certain details in order to focus on others. The challenge is for the creators of deep maps to help users understand the source of specific texts or images so they can assess them against their own evidential needs.

26 Mia Ridge, "There's a New Viewer for Digitised Items in the British Library's Collections." https://blogs.bl.uk/digital-scholarship/2016/12/new-viewer-digitised-collections-british-library.html.

27 Project Gutenberg (www.gutenberg.org/) is an important exception as texts are checked by volunteer "Distributed Proofreaders."

28 Ian Gregory, et al., "From Digital Resources to Historical Scholarship with the British Library 19th Century Newspaper Collection." www.researchgate.net/publication/297480687_From_digital_resources_to_historical_scholarship_with_the_British_Library_19th_Century_Newspaper_Collection.

29 This issue is not unique to the British Library. Naming "online resources" so that users can easily find them still seems difficult for CHIs.

30 Christopher Fleet, Kimberly Kowal, and Petr Přidal, "Georeferencer: Crowdsourced Georeferencing for Map Library Collections." www.dlib.org/dlib/november12/fleet/11fleet.html.

31 British Library LibCrowds Project, "LibCrowds Spring Newsletter." https://us11.campaign-archive.com/?u=08e409d3d85876a17ac4c1d09&id=3484f0e816.

32 British Library, "Georeferencing." www.bl.uk/projects/georeferencer.

10 The inexactitude of science

Deep mapping and scholarship

John Corrigan

In that Empire, the Art of Cartography attained such Perfection that the map of a single Province occupied the entirety of a City, and the map of the Empire, the entirety of a Province. In time, those Unconscionable Maps no longer satisfied, and the Cartographers Guilds struck a Map of the Empire whose size was that of the Empire, and which coincided point for point with it. The following Generations, who were not so fond of the Study of Cartography as their Forebears had been, saw that that vast Map was Useless, and not without some Pitilessness was it, that they delivered it up to the Inclemencies of Sun and Winters. In the Deserts of the West, still today, there are Tattered Ruins of that Map, inhabited by Animals and Beggars; in all the Land there is no other Relic of the Disciplines of Geography—purportedly from Suárez Miranda, *Travels of Prudent Men*, Book Four, Ch. XLV, Lérida, 1658.

Jorge Luis Borges, "On Exactitude in Science"[1]

Deep mapping and assemblages

The exact map of the Empire, described by the imaginary Suárez Miranda in Borges's microstory, despite its colossal dimensions and luxuriant detail was not a deep map. A full-scale *tracing* of territory, it was a useless cartographic conceit, its scientific exactitude its downfall.[2] As a relic of misplaced ambition it delivered neither practical serviceability as an artifact nor a spatial paradigm to be exploited in the interest of aggregating knowledge. It lacked, especially, the features that make a deep map valuable: active liaison with other knowledge, recognition of the complex multilayering of territory, and facilitation of navigation through space that often is mutable, inconsistent, and contradictory. Distinguished from the map of the Empire, a deep map is engaging because of its inexactitude.

One way of conceptualizing a deep map is to think of it as an assemblage. That term, assemblage, conjures the perspective of the philosophers Giles Deleuze and Felix Guattari. For them, "territory" is always fluid and dynamic, constituted by the shifting interrelationships of multiple heterogeneous components. As Brent Adkins remarks in his commentary on their ideas, territory is the site of "the interconnection of wildly diverse things."[3] Those things ceaselessly connect, disconnect, and reconnect in new ways, so that territory is ever shifting: territorialization, deterritorialization, and reterritorialization. A *map* of territory

DOI: 10.4324/9780367743840-10

for Deleuze and Guattari accordingly is "oriented toward an experimentation." It is not an intrinsic tracing of reproducible structures, vertical hierarchies, or linear genealogies. It is about multiplicities not binaries, mutualities not closed narratives. Objects, feelings, language, signs, power, body, geography: all and more are dynamically and intermediately assembled in a map. Mapping as such is a performance. As an assemblage, a map is largely about emergence rather than immanence.[4]

Deep mapping is central to the coalescing project of the spatial humanities. It is promising, complicated, frustrating, and evolving. It exemplifies the strengths and weaknesses that are at the heart of the spatial humanities. To a certain extent, a deep mapping is an observation of territory in the way described by Deleuze and Guattari, and likeminded researchers in the humanities and arts, who find intellectual investment in performance, multiplicity, and open-endedness to be useful. Conceptualizing a deep map as a fluid assemblage resonates with theoretical discourses ascendant in the humanities. Not incidentally, that resonance certifies deep mapping as a humanities constituent entitled to a place at the debating table with other humanities players. Deep mapping abets interpretation. It advances the broad agenda of humanities scholarship.

Deep mapping in as much as it is a form of scholarly practice in the spatial humanities, however, also draws upon scientific approaches to space that foreground quantification,[5] immanence (this is the real world and we are in it), structural stability and predictability (the limbs of the tree always grow from the trunk), and linear argument (that ends in "findings"). The elements of territory add up to something. There is a story to tell, a point to be made. A map, as Denis Wood (author of Chapter 2 in this book) has written elsewhere, makes an argument.[6] A deep map, as an argument, wades into debates, taking other maps as arguments and responding to them. In some cases, it may be that a deep map accomplishes that heavy-handedly, by attempting a wholesale decertification of previous observations of territory through challenges to traditional thinking about what maps are and what they do. Some maps that proffer a monoptical vision of space, of placial identities as firmly grounded and reliably consistent over time, are susceptible to the implicit critique of the deep map that displays identities as slippery and narratives as multivocal and provisional.

For example, *Mapping Emotions in Victorian London*, a site created by the Center for Spatial and Textual Analysis (CESTA) at Stanford, integrates multiple layers of testimony in mapping the emotional valence of various sites throughout the city. Like Rachel Smith's bicycling map of Berlin, and the recent work of Daniele Quercia, which map "happy" travel routes in urban settings, it is a crowdsourced map of feeling. Like Ian Gregory's geographic survey of feeling in Wordsworth and Gray's Lake District references, it engages historic literature. In joining those two approaches, it creates a multidimensional and complex field of data. Participants can engage excerpts from Victorian fiction in which places in the city are mentioned and then register their input about whether that place is happy, fearful, or unemotional. The large number of participants and the many different literary works made available ensure that many voices will be heard in the process of mapping emotion. Those voices will be temporally and spatially differentiated. Victorian authors and twenty-first-century readers, Londoners, and non-Londoners, those who have participated and those whose anticipated participation will affect the map in the future, together will draw the map. The collected data evidence frequent disagreements among parties about whether sites are fearful or happy. Questions arise about why fearful sites are directly adjacent to happy ones, or why a site is fearful at one point in time (there is a time slider on the page) and happy at another, or why proximate sites over time can fluctuate from a state of opposed emotions to happy together. The emotional map of London that results is not a finished drawing but a continuously changing assemblage. *Mapping Emotions in Victorian London* consequently is a relatively big data project that makes different kinds of arguments, some of which do not line up in ways commensurate with the logics of traditional mappings. That might be why the site announces itself with what in effect is a caveat: "a crowdsourcing project designed to expand possibilities for research in the humanities."[7] That undersells its importance. It is a resource, but it is an argument as well. It argues that a deep map, apart from whatever specifics it can provide about a site (e.g., Covent Garden *is* a happy place), is itself a trial balloon in defining what we mean by territory. It is an argument for variability, change, and unevenness.

If the Victorian emotions site is an experiment in thinking about territory, it is an experiment that is still running. Unlike conventional scientific experiments that have termini, and that prove or disprove a clearly defined thesis (or fail to do both, yet still "end"), deep mappings tend to begin with more general aims, which are sometimes so distensible that there is no way to know when the experiment is over. That is not the case with all spatial humanities projects. Many are tightly focused and deliver conclusions. Some of the other publications in this Spatial Humanities books series make precise and compelling arguments about all kinds of things. *Troubled Geographies: A Spatial History of Religion and Society in Ireland*, for example, clearly delineates the manner in which religious communities were constituted along economic and class lines, how the geography of the potato blight played a central role in that complex, and how partition both reinforced and undermined socio-religious formations.[8] Similarly, the case studies in *Geographies of the Holocaust* strikingly demonstrate how the Holocaust was a multiplanar geographic reality. But deep mappings such as *Victorian London*, as a specific initiative of the spatial humanities, are open-ended. The question then arises as to whether they make an argument about a specific subject if they never conclude.

Deep mapping and science

Deep mapping has one foot in science and one foot in the humanities. There are several ways in which to recognize the associations with science, none of which require lengthy discussion of what science is, philosophically or historically. Most obviously, deep mapping enjoys rapprochement with the advance guard of digital technologies, and in fully engaging them it inevitably takes on some of the epistemological baggage redolent in them. Deep mapping often invokes quantitative methods, ranging from the depiction of space in coefficients of latitude and longitude, to calculations of spatially defined demographic percentages, to word counts. Deep maps often frame interpretation as a process built from data more so than from theory. Deep maps, however, do not as ambitiously seek reductions as positivistic science does, nor do they assume a separation between the knower and the known. They are post-positivistic in a way that partially aligns them with the paradigm outlined by Thomas Kuhn and Karl Popper.[9] In general if we think of science as post-positivistic there is room for deep mapping under that conceptual umbrella. The ends of post-positivistic scientific research overlap in several ways with the aims of deep mapping.

For perspective, it is useful to consider how a deep map might compare to scholarship in the sciences. Scientists build experiments out of their knowledge of their fields. They run those experiments to advance the frontiers of those fields and to test the adequacies of the theories that underpin them. Nevertheless, however much individual experiments come to conclusions, the Big Experiment never concludes. Science, as philosopher Karl Popper and his followers have asserted, is always a question about itself. "The game of science is, in principle, without end," wrote Popper. "He who decides one day that scientific statements do not call for any further test, and that they can be regarded as finally verified, retires from the game."[10] Individual research papers make arguments, but in the broader view—at least for the Popperians—such papers are always framed by expectations that some previous research will be challenged and overthrown. It can be argued that for all their empirical soundness and logic, scientific arguments, in the long view, equally are intermediate, fluid, and open-ended. Some writers, reflecting on the impact of big data capabilities on scientific research, have extended that observation to characterize current science as an enterprise in transition. Hey and Tolle, for example, have proposed that data-intensive science, where statistical exploration and data mining on a massive scale constitute research, in fact is entering a "fourth paradigm" of "exploratory science." That emergent paradigm is conceptualized as the maturing of science beyond previous stages of experiment (empiricism), theory (modeling), and computation (complex simulations). Some scholars have been especially clear in identifying the ways in which the turn to big data has been involved in the shift.[11]

A deep map as an assemblage is a different kind of research than "exploratory science." However much we might concede that science is open-ended in the long term, and that in this season of big data a paradigm of scholarship-as-discovery is ascendant, the fact remains that a Deleuzittarian assemblage, even if it is an "experimentation," cannot be construed to be in direct conversation with relevant subfield-defining scholarship. There is background, as previously noted, but it is at the level of meta-analysis, an engagement with the very notion of "territory" and "map." A "map," an assemblage for instance of a wasp pollinating an orchid (an example offered in *A Thousand Plateaus*) tells us little or nothing about why the wasp does that, how the orchid collaborates beyond the instant of the event, or how that occurrence changes the territory other than that it amounts philosophically to a reterritorialization, an emergence, "a becoming-wasp of the orchid and a becoming-orchid of the wasp."[12] There is an aesthetic aspect to imagining territory in such a way, and in that sense there is ground for a potential collaboration with humanities scholarship, which frequently is attentive to aesthetics, and particularly in the crafting of a prose explication of an interpretation. And like lucid language in a book, beautiful graphics and images on a deep map website contribute to an argument.

However, aesthetics are not enough if the point is to participate in a scholarly collective that debates not only what is meant by the terms territory and map but also how and why the wasp and orchid came to occupy that territory together. Those latter questions reference histories and genealogies, interconnected conversations among specialists, and additional related data. They position a map within the intellectual cultures of the humanities and social sciences. The situatedness of the academic deep mapper within the academic collective defines the act of mapping in a way that implicates broader issues involving the territories to be mapped. The scholarly collective—including its discourses, disciplines, epistemologies, archives, and associations—situates the deep map. The question consequently has to do with how deep maps surface the conditions of their situatedness, where conditions refer to debates within the field and other disciplinary background. Deep mapping as a spatial humanities enterprise however so much it shares in the project of the assemblage—the instability, aesthetics, and diversity of the assemblage—represents shared, often field-specific assumptions cultivated within the humanities. In other words, a deep map, for all of its slipperiness and effort to polyopticize territory, is constrained by its creator's agendas. All of this defining of deep mapping as discursively framed, which is just an instancing of critical theory that is well rehearsed over a full generation, suggests something relatively simple: It might be that we have to work harder to understand exactly what sort of scholarly interventions a deep map offers. We should look to the disciplinary—or, as in most cases, interdisciplinary—embeddedness as well as to the aesthetics and the paradigm-challenging approach.

As noted, deep maps differ from what we might call "thin" maps in that they acknowledge differences of viewpoint and are structured to accommodate them. There are surprises in deep maps. As open-ended experiments, perhaps heralding "exploratory science," but equally as deliberate creations of humanities that cleave to interpretation as its rationale, they can yield unexpected views of territory. They are different and hard to categorize. This, as previously stated, is one of their weaknesses, but it does not delegitimate them as scholarship.

As a member of a dissertation committee in brain science I recently attended the defense. The study was not able to build effectively on what was strong pilot data and the statistical work, which was sophisticated, did not show the patterns that the investigator had hoped. The research proved inconclusive, or, to be more precise, the experiment did not prove what it set out to prove. Nevertheless, it was agreed that the dissertation was high quality in terms of its rigor, and it passed unanimously. There were many surprises that arose from the exercise, some of which offered auspicious leads for continued research on the topic. Others pointed to equally promising ways to rethink some of the central field concepts within the discipline. In short, the dissertation was good scholarship in spite of the fact that it resulted not in a proof but in a document that prompted new questions, could serve as a resource—by virtue of its well-collected data and exemplary statistical work—and displayed a failed argument. The committee's approval was not, I think, explicitly grounded in an embrace of what Rob Kitchin has termed "Big Data Analytics," which he defines as "an entirely new epistemological approach for making sense of the world; rather than testing a theory by analysing relevant data, new data analytics seek to gain insights 'born from the data.'" That data-driven science on the whole is:

> more open to using a hybrid combination of abductive, inductive, and deductive approaches to advance the understanding of a phenomenon. It differs from the traditional, experimental deductive design in that it seeks to generate hypotheses and insights "born from the data" rather than "born from the theory."[13]

But the committee did strongly acknowledge the process of the dissertation to be sound, the several auxiliary theories—which were actually "born from the data"—to be worth pursuing, and the data itself rich and plentiful.[14]

Seven aspects of deep mapping scholarship

Revisiting the theme of deep mapping as argumentation, it is possible, against a contextualizing background of some aspects of science, aesthetics, and philosophy, to speak of several different ways in which a deep map is an argument. Deep mapping in the humanities areas (and in cognate social sciences fields) first of all challenges the adequacy of previous kinds of mapping. A deep map by virtue of its construction argues that the business of the humanities is to investigate the sufficiency of commonly accepted understandings of territory and map, node and network, text, and space. Some deep maps are more ambitious in that enterprise than others. But many deep maps will disclose something about space that challenges how we think about it. A deep map can alter our conceptualization of territory by showing it to be more racialized or gendered than we had thought. It can demonstrate the relevance of natural topographic features or built environments in ways that disturb previous understandings which took such features for granted. It can show territory to be ambiguous, or in some cases to be flatter, less variable, and more inert than previous mappings had proposed. A deep map can argue against the relevance of categories that conditioned the construction of previous maps. In its susceptibility to interrogation and user-directed exploration it can manifest as the shifting depiction of a territory that in the end might be better known to the user than to the creator. In a recent essay, Bruno Latour and Valérie November ask why the "base map" is typically understood to be a geographic drawing and how other maps are then superficially layered on top of that base map. Perhaps, they propose, the base map could be, for example, a map of risk, or some other non-geographic layer. For Latour and November, treating the digital map as a living artifact characterized essentially by communicativity, navigability, and reflexivity discloses it to be *practical activity* rather than mimesis. A deep map as a digital creation is for navigation, and the map changes as the navigator's needs and interests change. That is an argument about what a map is, what is meant by territory, and the role of "navigators" in making and using the map.[15]

Second, deep maps argue that territory can be read in different ways, and often they invite contradictory interpretations. Shelly Fishkin Fisher, in outlining an American Studies digital literary project, discusses how a deep map as an assemblage of texts is like a palimpsest. What has been overlaid nevertheless shows through, disrupting unilayered readings and "sometimes reflecting conflicting interpretations of the material involved."[16] What she describes in characterizing textual territory applies as well to broader deep mapping projects, including those that do begin with a geographic base map. Depending on the layers, the depth and quality of the data, and the responsiveness of the deep map to user interests, conflicting interpretations easily might arise. Such conflicts then complicate the status of a deep map itself as an argument. But like scientific experiments that sometimes end in findings that support differing interpretations, a deep map is a scholarly product. Its scholarship in such a case is not located in its capability to mount a direct challenge to a specific interpretation, but rather in its capability to generate competing interpretations or auxiliary theories.

Third, deep maps sometimes can respond directly to previous interpretations. The London emotions site offers possibilities for users, as they navigate the site, to construct interpretations that respond to issues considered by them to be important in the subfield. But it is possible as well to take the project itself, overall, as an argument favoring the expression and variety of emotional life in Victorian London. That is, it can contest what in the past sometimes was characterized as the blunted effect of Victorian manners, the obsessive attendance upon "emotional management," the conformity to repressive feeling rules, and the "secret life" of feeling.[17] The site also could be credited with making an argument for the relevance of specific locations for various feelings in London, especially with regard to public places. Because the site makes those points through its curation of literary traces it would comport with what scholars in the digital humanities as well as artists often claim in defining curation as an argument.[18]

For some, the question arises as to whether curation qualifies as scholarship in the humanities, and there has been much recent discussion of the issue about the digital humanities as whole. That conversation is complicated, but it might be simplified by considering a comparison between an edited volume of papers—a historic figure's correspondence, perhaps—and a well-curated, visually robust website. Both require knowledge and insight and a level of skill in making choices about how to select, arrange, and annotate the various items. Edited volumes of primary sources have been recognized as important scholarship for generations, and such collections have been reviewed for their display of technical competencies as well as their capability to demonstrate new interpretations of the sources.[19] Deep maps can make an argument like some edited hard-copy volumes through their capability to frame a diverse range of objects in a coherent manner and in doing that prompt focused thinking about the meaning of that archive for disciplines relevant to it.

Fourth, deep maps are a collaborative scholarship. By this I do not refer exclusively to the collaboration between the official creators of a deep map.[20] Rather, I describe deep mapping as collaborative scholarship that often involves more than the instigating team, and the outcomes of deep maps, which are shaped in part through the contributions of subsequent participants, as collaborative arguments. There are ambiguities in this kind of scholarship. To include the user of a spatial humanities website as a member of a team that produces scholarship admittedly is to envision a model of scholarship that abrades familiar notions of authorship. But a crowdsourced deep map typically includes many diverse authors. Each of the participants in the London emotions site, whether they add a text, an annotation, or an opinion about the kind of feelings associated with a location, are in some measure collaborators. The quality of the participation matters, and deep mapping that invites crowd sourcing has not developed broadly accepted criteria of quality that might apply across a range of projects. They are not alone in that, however. A physics paper published in *Physical Review Letters* in 2015 numbered 5,154 authors. The first nine pages of the paper described the research. The remaining 24 pages listed the authors.[21] A deep map that is crowdsourced and includes several thousand individual contributions will feature some contributions that will be more important than others. But the collective will move the deep map in certain directions, as some contributions coalesce around a center or multiple, even competing centers.

The collaborative process in deep mapping ventures that engenders the emergence of multiple centers, which are joined in an overall network, might be understood with reference to November's and Latour's insight that a map is a collection of "signposts" and that when a person engages a map she moves from one signpost to another. "Maps," they write, "should be considered dashboards of a calculation interface that allows one to pinpoint successive signposts while moving through the world, the famous *multiverse* of William James."[22] Those signposts can change in terms of their relevance to other signposts. Each distinctive signpost becomes more or less important as it is encountered, depending on what has come before it and where the navigator wishes to go. Thus, there is a pluralism to a map, rather than an absolute unity. In a collaborative map there are many of what the philosopher William James called "small systems,"[23] which were for him also "small communities."[24] That pluralism comports with James's own claim about the "multiverse" that that "there is no point of view, no focus of information extant, from which the entire content of the universe is visible at once."[25] One navigates from one signpost to another, again, depending on circumstances. The relations between constituent parts of the broadly collaborative deep map are similar to that of the elements of the multiverse, so that:

> "Our multiverse still makes a universe; for every part, tho it may not be in actual or immediate connexion, is nevertheless in some possible or mediated connexion, with every other part however remote, through the fact that each part hangs together with its very next neighbors in inextricable interfusion," as "small systems" which together form the whole.[26]

Deep mapping as collaborative scholarship can display signposts, create "small systems" that are related in a fluid connection to other parts of the map. Again, the idea of authorship redolent in this model is unfamiliar, but it is authorship nevertheless, when understood from the perspective of argumentation as a plural endeavor.

Fifth, deep mapping is scholarship that explores the application of technology and in some cases comprises ambitious interpretations of it. The humanities are in transition as the digital world progressively influences research and publication. Scholars increasingly adapt technology to serve their research, and in so doing position themselves to contribute both within their fields and to a broader advance of technological applications. Oftentimes the two cannot be separated, as the interpretations that arise from enhanced technological utilization are fully understandable only within the epistemological framework of a technologically enabled analysis. A useful analogy in terms of the recent course of humanities research is the reemergence[27] of what was known during its peak in the 1970s as chart-and-graph history and currently is represented by journals such as *Cliodynamics: The Journal of Theoretical and Mathematical History*. Early on, it availed itself of what was once called "humanities computing," and especially in its forays into the area of economic history. Its exploitation of computing technology has kept pace with the analytical objectives of its practitioners, and the joining of those two aspects of scholarship is reflected in its coalescing research paradigm. According to Peter Turchin, one of its promoters, it is an approach requiring scholars to "collect quantitative data, construct general explanations and test them empirically on all the data rather than on instances carefully selected to prove our pet narratives."[28] While not all deep maps are constructed in order to provide general explanations, they are, as noted earlier, experiments which, in incorporating technology, are shaped in certain ways by the epistemologies redolent in the technology. Deep maps break ground for interpretation in that regard. As experiments with technology—some of which result in interpretative leverage and some of which remain open-ended explorations—they are part of a movement across the humanities and social sciences that collectively is defining the applicability of digital and other technologies to scholarly argumentation.

Sixth, deep maps often are public scholarship. Because of their reliance in most cases on digital technology, including not only software that aids in collection and analysis but also the digital platforms that make research available online, many deep maps result in interactive projects accessible to nonspecialists. Public scholarship is neither less rigorous nor less pointed than highly technical research. It is a form of publication that presents data and interpretations in ways that avoid jargon and do not levy the requirement of extensive field-specific knowledge upon the reader. Deep maps that can be easily accessed online, and that are constructed in ways that enable not only intellectual accessibility but also public participation, combine research with pedagogy. Their value as public scholarship lies precisely in their capability to manage that relationship. An example of how a spatial humanities website can multitask as research and pedagogy is *Danteworlds*[29] which invites participants to navigate through the territories of Hell, Purgatory, and Paradise. Along the way, people, texts, ideas, and art populate the landscapes and, in many cases, can be explored in detail. The site is a model teaching project but at the same time the rich traveler's guides for the site were published in two books from the University of Chicago Press.[30]

Seventh, there are aesthetic dimensions to deep maps. Deep maps must be imagined just as any argument in any humanities field has to be imagined. They must also be designed in ways that typical scholarship in the humanities is not. There are considerations of representation and articulation involving scale, structure, visual cues and codes, color, the interrelationship of text and images, and other aspects of composition that are either absent from or largely diminished in traditional humanities scholarship. If a deep map is to be constructed on a digital platform, the requirement of quality demands even more stringent controls, careful foresight, and effective configuration. There is art to a deep map, and it is crucial to its success.

Via media: flexibility in assessing deep maps

One way to assess scholarship is to estimate its impact. Institutions of higher education have developed various ways in which to "measure" the impact of scholarship produced by their faculty and to estimate the potential impact of research by students. The latter effort is mostly limited to the judgment of committees of field experts who advise or otherwise oversee student work, most importantly doctoral research. Such assessments rarely rely on metrics beyond the vote of the dissertation committee because students generally have spare publication records if they have them at all. Instead, such assessments proceed through the deliberation of informed experts who collectively judge the quality of student work. There is an effort to gauge quality, rather than quantity, an effort to predict impact rather than to confirm it. Universities award doctoral degrees on that basis. The determination of impact of faculty research in the sciences and social sciences, on the other hand, while also attempting a judgment of quality, typically is titled toward impact measured by references in citation indices, publication counts, and other quantitative means. That approach coalesced within the natural sciences, where the predominant genre of publication is the article (frequently multiauthored) that customarily includes exhaustive references to previous works that likewise were published as multiauthored articles. The impact of scholarship in such a case is a quantity. The shortcomings to this method have been well documented and include the negative steering effects of indicators—the funneling of research into narrow and "spectacular" projects and away from marginal topics.[31]

In between those two ways of assessing the impact of research (e.g., dissertation/ science publications) is the practice in the humanities. While humanities scholars do write articles (generally single-authored), and assessment committees do count them, the critical publication in most humanities fields is the book, and most humanities scholars devote their time to researching and writing books. A few good books can establish a scholar's stature. But books are not assessed by counting citations. Even some career-defining Pulitzer-winning history books accumulate less than 400 citations over a period of 10 years. The impact of a book in the humanities is measured more in terms of the quality of its impact. Humanities books that are of high-quality change the ways scholars in a field conceptualize their core research issues. In attempting to measure quality, humanities scholars have experimented with a wide range of tools. The lesson, so far, is that assessment of scholarship in the humanities sometimes relies too much on tacit knowing rather than explicit knowledge, but that it is a flexible process.[32]

It is the tradition of flexibility in thinking about humanities scholarship that has both fostered the emergence of deep mapping as a humanities research agenda and maintains it. The seven different ways in which deep mapping is scholarship outlined previously are not stand-alone criteria. A deep map might exemplify quality in several different areas at once, or even across all seven. Assessments of the quality of deep mapping scholarship must appreciate the ways in which the contributions of a deep map often are manifold and overlapping. Deep maps are assemblages, and as such are active participants in the craft of humanities interpretation, as metalevel interventions that challenge familiar categories of analysis, as pedagogical forays that make knowledge accessible, and in many other ways or combinations of ways. They also, to a certain extent, are scientific experiments, arranging data, interrelating it, arguing for a certain understanding of it, and responding in course to previous interpretations of it. Deep maps are a *via media*, persuading "through media" and constituting a middle way between the humanities and sciences. That status, full of promise but not yet fully actualized, is part of the reason for the surge of scholarly interest in deep maps. A deep map is an original form of scholarship that surfaces disciplinary and philosophical tensions that lie within the humanities and exploits them in creative and productive ways.

Notes

1 Jorge Luis Borges, "On Exactitude in Science," in Jorge Luis Borges, *Collected Fictions*, trans. Andrew Hurley (New York: Viking, 1998), 325. Originally published within a longer writing, "Museo," in *Los Anales de Buenos Aires* 1 (1946) under the name B. Lynch Davis, a mutual pseudonym of Borges and Adolfo Bioy Casares.

2 Useful perspective is in Alfred Korzybski, *Science and Sanity: An Introduction to Non-Aristotelian Systems and General Semantics*, 5th ed., second printing (Brooklyn: Institute of General Semantics, 1994), 1933. "A map is not the territory it represents, but, if correct, it has a similar structure to the territory, which accounts for its usefulness. . . . If we reflect upon our languages, we find that at best they must be considered only as maps; a word is not the perfect object it represents; and languages exhibit also this peculiar self-reflexiveness, that we can analyse language by linguistic means. This self-reflexiveness of language introduces serious complexities, which can be solved only by the theory of multiordinality" (p. 58).

3 Gilles Deleuze and Felix Guattari, *A Thousand Plateaus: Capitalism and Schizophrenia*, trans. Brian Massumi (London: Continuum, 1987); Brent Adkins, *Deleuze and Guattari's a Thousand Plateaus: A Critical Introduction and Guide* (Edinburgh: Edinburgh University Press, 2015), 24.

4 For discussion of the "rhizome" as a representation of mapping as assemblage see *A Thousand Plateaus*, 12. A useful introduction to thinking about space in Deleuzian

terms, written by a philosopher and a geographer, is Mark Bonta and John Protevi, *Deleuze and Geophilosophy: A Guide and Glossary* (Edinburgh: Edinburgh University Press, 2004).

5 Deep mapping, even in a Deleuzian key, has something in common with dynamical systems theory and chaos theory in mathematics. For the compatibility of Deleuzian ideas with mathematics see John Johnston, *The Allure of Machinic Life: Cybernetics, Artificial Life, and the New AI* (Cambridge: MIT Press, 2008). See also James Williams, *Gilles Deleuze, Difference and Repetition: A Critical Introduction and Guide* (Edinburgh: Edinburgh University Press, 2013); Brian Massumi, *A User's Guide to Capitalism and Schizophrenia: Deviations from Deleuze and Guattari* (Cambridge: MIT Press, 1992); and Manuel DeLanda, *Intensive Science and Virtual Philosophy* (London: Bloomsbury Academic, 2002).

6 Denis Wood, John Fels, and John Krygier, *Rethinking the Power of Maps* (New York: Guilford Press, 2010), 42–4.

7 Daniele Quercia, Rossano Shifanella, and Luca Maria Aiello, "The Shortest Path to Happiness: Recommending Beautiful, Quiet, and Happy Routes in the City," Proceedings of the Conference on Hypertext and the Social Media (Hypertext), 2014. www. di.unito.it/~schifane/papers/hypertext14 shortest.pdf (accessed July 21, 2017); Rachel Smith, "Dynamic Connections: A Crowdsourced Bike Map of Information for Cyclists in Berlin," 2012. www.bmwguggenheimlab.org/where-is-the-lab/berlin-lab/berlin-lab-city-projects/dynamic-connections-map; Ian Gregory, "Mapping the Lakes: A Literary GIS." www.lancaster.ac.uk/mappingthelakes/; *Mapping Emotions in Victorian London.* https://about.historypin.org/2015/04/14/mapping-emotions-in-victorian-london/. See also Patricia Murrieta-Flores, Christopher Donaldson, and Ian Gregory, "GIS and Literary History: Advancing Digital Humanities Research through the Spatial Analysis of Historical Travel Writing and Topographical Literature," *Digital Humanities Quarterly* 11 (2017). www.digitalhumanities.org/dhq/vol/11/1/000283/000283.html.

8 Ian N. Gregory, Niall A. Cunningham, C.D. Lloyd, Ian G. Shuttleworth, and Paul S. Ell, *Troubled Geographies: A Spatial History of Religion and Society in Ireland* (Bloomington, IN: Indiana University Press, 2013).

9 Karl Popper, *The Logic of Scientific Discovery* (London: Routledge Classics, 2002 [*Logik der Forschung*, 1934]) and *Conjectures and Refutations: The Growth of Scientific Knowledge* (London: Routledge Classics, 2002) [1963]); Thomas S. Kuhn, *The Structure of Scientific Revolutions* (Chicago, IL: The University of Chicago Press, 1962).

10 Popper, *The Logic of Scientific Discovery*, 32.

11 T. Hey, S. Tansley, and K. Tolle, "Jim Grey on eScience: A Transformed Scientific Method," in T. Hey, S. Tansley, and K. Tolle, eds., *The Fourth Paradigm: Data-Intensive Scientific Discovery* (Redmond, WA: Microsoft Research, 2009), xvii–xxxi. A discussion of the ways in which big data research has been adopted into the sciences, social sciences, and humanities is Rob Kitchin, "Big Data, New Epistemologies and Paradigm Shifts," *Big Data and Society* 1 (April 2014). DOI: 10.1177/2053951714528481.

12 See Deleuze and Guattari, *A Thousand Plateaus*, 11–13.

13 Kitchin, "Big Data, New Epistemologies and Paradigm Shifts," 2, 5–6. https://journals. sagepub.com/doi/full/10.1177/2053951714528481.

14 The dissertation I reference is: Matthew Michaels, *Gender Norms, Sex, Sexual Orientation, and Suicidal Behavior* (Ph.D. dissertation, Florida State University, 2017).

15 Valerie November and Bruno Latour, "Entering a Risky Territory: Space in the Age of Digital Navigation," *Environment and Planning D: Society and Space* 28 (2010): 581–99.

16 Shelley Fishkin Fisher, "'Deep Maps': A Brief for Digital Palimpsest Mapping Projects (DPMPs, or 'Deep Maps')," *Journal of Transnational American Studies* 3 (2011): 1–31.

17 Some of these issues are discussed in Anne-Marie Millim, *The Victorian Diary: Authorship and Emotional Labour* (Aldershot and Burlington, VT: Ashgate, 2013); Thomas Dixon, *Weeping Brittannia* (Oxford: Oxford University Press, 2015); Rachel

Ablow, ed., *Affective Experience and Victorian Literature* (Ann Arbor: University of Michigan Press, 2010).

18 Daniel Price, Rex Koontz, and Lauren Lovings, "Curating Digital Spaces, Making Visual Arguments: A Case Study in New Media Presentations of Ancient Objects," *Digital Humanities Quarterly* 7 (2013).

19 Trevor Muñoz proposes that curation is, first of all "publishing," sidestepping the thornier problem of scholarship per se (not all that is published is scholarship). Nevertheless, leveraging Robert Darnton's understanding of the history of books, he states that "taking 'publishing' as a category to be re-imagined rather than a pre-existing workflow to stepped into, . . . this is data curation-as-publishing" ("Data Curation as Publishing for the Digital Humanities," *Digital Humanities Quarterly* 2 (2013).

20 Some deep maps are not crowdsourced, and so the assessment of their scholarship as collaborative works would proceed differently. Many nevertheless are created by teams of scholars rather than individuals.

21 G. Aad, et al., "Combined Measurement of the Higgs Boson Mass in pp Collisions at $\sqrt{s}$= 7 and 8 TeV with the ATLAS and CMS Experiments," *Physical Review Letters* 114 (2015): 191803.

22 November and Latour, "Entering a Risky Territory," 581. The world they describe is, they say, the "multiverse" of William James.

23 William James, *Pragmatism*, ed. Frederick H. Burckhardt, Fredson Bowers, and Ignas K. Skrupskelis (Cambridge, MA: Harvard University Press, 1975), 67.

24 Deborah J. Coon, "One Moment in the World's Salvation: Anarchism and the Radicalization of William James Author(s): Source," *The Journal of American History* 83, no. 1 (June 1996): 70–99. See especially p. 83 on small systems/small communities.

25 James, *Pragmatism*, 72.

26 James, *Pragmatism*, 67.

27 Marc Perry, "Quantitative History Makes a Comeback: Historians Argue for a Scientific Study of the Past," *Chronicle of Higher Education*, February 25, 2013. www.chronicle.com/article/Quantifying-the-Past/137419/.

28 Peter Turchin, "Arise 'Cliodynamics'," *Nature* 454 (2008): 34–5. DOI: 10.1038/454034a; Published online 2 July 2008.

29 https://danteworlds.laits.utexas.edu/.

30 Guy Raffa, *Danteworlds: A Reader's Guide to the Inferno* (Chicago, IL: The University of Chicago Press, 2007) and *The Complete Danteworlds: A Reader's Guide to the Divine Comedy* (Chicago, IL: The University of Chicago Press, 2009).

31 D. Fisher, K. Rubenson, K. Rockwell, G. Grosjean, and J. Atkinson-Grosjean, *Performance Indicators and the Humanities and Social Sciences* (Vancouver, BC: Centre for Policy Studies in Higher Education and Training, 2000); Martin Hose, "Glanz und Elend der Zahl," in Claudia Prinz and Rüdiger Hohls, eds., *Qualitätsmessung, Evaluation, Forschungsrating. Risiken und Chancen für die Geschichtswissenschaft?* (Berlin: Historisches Forum, 2009), 91–8. This is an online publication: https://edoc.hu-berlin.de/bitstream/handle/18452/18471/HistFor_12-2009.pdf?sequence=1&isAllowed=y.

32 Michael Ochsner, Sven E. Hug, and Hans-Dieter Daniel, "Humanities Scholars' Conceptions of Research Quality," Research Assessment in the Humanities: Towards Criteria and Procedures, Springer International Publishing, April 29, 2016, 43–69. https://link.springer.com/chapter/10.1007/978-3-319-29016-4_5.

11 Convergence

GIScience turns to the spatial humanities

Karen Kemp

Editors' Note: *Making Deep Maps* is the latest book in a three-volume set—the earlier titles are *The Spatial Humanities: GIS and the Future of Humanities Scholarship* (Indiana University Press, 2010) and *Deep Maps and Spatial Narratives* (Indiana University Press, 2015)—that defines the new field of spatial humanities and explores its methods, applications, and innovations. The editors of these three books asked Karen Kemp, a well-known GIScientist who has long worked at the intersection of geography and the humanities, to reflect on the contributions and still-emerging potential of spatial humanities and its most innovative form, deep mapping. A participant to the 2008 workshop that shaped the field in its first expression and to the 2010 book, Kemp offers both a personal and professional coda to the three volumes collectively and points to the growing influence of spatial humanities on interdisciplinary scholarship.

A series of three NEH Advanced Institutes, held in 2008, 2012, and 2018 at the Polis Center in Indianapolis and West Virginia University Morgantown led to the crafting of the three books in this series on The Spatial Humanities. The first two—*The Spatial Humanities: GIS and the Future of Humanities Scholarship* (2010) and *Deep Maps and Spatial Narratives* (2015)—helped to define a new field and introduced deep mapping as an approach to problems that humanists encountered as they sought to apply GIS and geographical concepts to problems of interest. With this volume, these collections of essays aptly record the evolution of the discussion over the 10 years. As a geographic information scientist interested in helping humanities scholars apply and adapt geographic information systems (GIS) to their own scholarship, I was fortunate to attend the first of these workshops. And now, more than a decade later, I was delighted when the editors of this series asked me to write a coda for this set of conversations. I am happy to report that there is clear evidence of great progress in this dialog.

When I attended that first expert workshop, I had already been working with scholars in the Electronic Cultural Atlas Initiative (ECAI) out of University of California Berkeley for several years. That is where I began enjoying illuminating conversations with the editors of these volumes about the role of mapping in the humanities. My intention, what I thought was my role in the group at the time, was to help humanities scholars not only understand how GIS might be used

DOI: 10.4324/9780367743840-11

(e.g., how to put their "data" into "tables" so we could map it), but also, more importantly, have deep conversations with them about what they really wanted and needed from a mapping tool in order that I could articulate to the GIScience community how GIS might evolve to meet their needs. In retrospect, it was a naive position to hold, and one that clearly frustrated the humanities scholars who really didn't want to put their "data" into "tables" or make spatially precise maps that followed our traditional cartographic rules.

It was at this first workshop that the role of deep maps in providing a place of communication between the spatial scientists and the humanities scholars emerged. My most important takeaway from that first meeting was that I didn't actually understand how humanists do their work. What is their process of scholarship? Satisfyingly, the books in this series, particularly the second book, articulate in rich detail the need for humanities scholarship to embrace "multiplicity, simultaneity, complexity, and subjectivity."[1] By confronting the uncomfortable positivism of GIS, rather than just rejecting it, these scholars have articulated their need "to replace this more limited quantitative representation of space with a view that emphasizes the intangible and socially constructed world and not simply the world that can be measured."[2] That is the kind of insight that I had been looking for in my early conversations.

What have these volumes revealed?

I did not attend the second and third workshops, but these volumes provide a view into the evolution of the conversation over the past decade. Let me highlight some key themes I see here.

The nature of scholarship in the humanities

There is much in these volumes to satisfy the GIScientist seeking insight into what humanities scholarship is, and about the uncertainties, imprecision, and multiplicity that drive it. There is also a challenge here for humanists to use the spatial humanities as an opportunity to rethink their scholarly storytelling methods, and to acknowledge how they intervene in the interpretation of literature, history, society, and place.

The discipline of Geography, often the home of GIScience at universities, has been divided for more than half a century between the primacy of the quantitative versus qualitative perspectives; between the notions of space versus place. The ascendancy of GIS has amplified that discord. In the early 1990s an impassioned critique of GIS emerged from the critical theory side of Geography which called into question many of the disconnects the scholars in these volumes have struggled with. For me, however, the eloquence of the scrutiny of humanities scholarship in these volumes is far more convincing. Here, it is not just about unrepresented communities and the subversive objectives of mapping organizations, it is about how engaging unmappable, post-positivist perspectives can provide understanding of the complex multifaceted world; how the voices of those we might not otherwise hear can be illuminated and how the readers themselves can be engaged as guides and contributors to the story being told.

There are many great "sound-bites" to help a reader understand the nature of humanities scholarship. Here are just a few of them.

On the role of place in humanities scholarship:

- Places are integrals of human experiences and should be represented as such.[3]
- The concept of place appeals to humanists, who see it as the carrier of culture, which implies that no one class or group controls it.[4]
- It is the contested constructedness of place—the linking of locale, events, and process that make up local place—and that of place-making and of the cultural practices employed for place-making, which intrigues humanists rather than the perhaps less subtle GIS emphasis on generalized space.[5]

On the process of humanities scholarship:

- Historians exercise selectivity, simultaneity, and shifting of scale in manipulation of space and time to construct narratives that interpret the past.[6]
- Humanities scholars like to draw a suite of conclusions at the same time, and not all of those conclusions might comport exactly with each other.[7]
- A humanities GIS-facilitated understanding of society and culture may ultimately make its contribution in this way, by embracing a new, reflexive epistemology that integrates the multiple voices, views, and memories of our past, allowing them to be seen and examined at various scales; by creating the simultaneous context that we accept as real but unobtainable by words alone; by reducing the distance between the observer and the observed; by permitting the past to be as dynamic and contingent as the present.[8]

On deep maps as a method for scholarship:

- Deep maps are intentionally subversive, imprecise, complex, reflexive, sumptuous, and resplendently untidy.[9]
- A deep map is an original form of scholarship that surfaces disciplinary and philosophical tensions that lie within the humanities and exploits them in creative and productive way.[10]
- Deep mapping can be seen as a way of emphasizing the role of place as well as that of space in humanistic studies and to incline GIS toward a greater empathy in experiencing the lived world rather than its current heavy focus on the spatial characteristics and mapping of the physical and tangible world.[11]

Deep maps are an artistic medium with a carefully crafted narrative

Stephen Robertson and Lincoln A. Mullen, in this volume, speak eloquently to the notion of a narrative driving the "reader's" experience of a deep map:

The aim of the deep map should be to translate for the past so that it can speak. What is needed is a way of navigating through that chaos in order to craft meaningful experiences, or experience the meaning of the past. And that act of translation, like much history, can come through narrative.[12]

Robertson and Mullen observe, however, that the presence of a narrative means the digital resources linked into the deep map cannot be random or haphazard. There must be a logic and a pathway the author wishes to be followed. A deep map is not enlightening without a narrative, or better, several narratives that can be explored.

Taking it to scale, problems of the digital resources

Deep maps as described in the final volume are magical artistic enterprises, one of a kind, mutable and impossible to crystalize. Creating a performative deep map is a work of creativity requiring careful attention to choices made in contents and connections. These are rare offerings and cannot be mass produced. Therein lies an important distinction. The emergence of a deep map is not the result of the presence of an appropriate tool (i.e., a deep map GIS) or a large but random collection of digital resources (such as the digital archives of the British Library described by Mia Ridge in her chapter) that can be linked into it. Like a humanities scholar's lifework in one or two volumes, so too is a scholarly deep map a culmination of years of effort. This suggests that deep maps will never become plentiful.

How the series has succeeded

There are several ways I believe this series is of significant value. Many chapters richly elucidate how place is more important than space to humanities scholars. This is key to understanding the spatial tool needs of these scholars. Many of the strengths and shortcomings of traditional GIS when used as a method for humanities scholarship are highlighted. Several chapters provide robust examples of digital efforts that are at least partially successful at addressing these shortcomings. These examples include ECAI's TimeMap, historian Ed Ayer's Valley of the Shadows Project, literary masterpieces such as those described by Dennis Wood, and Trevor Harris' explorations in immersive environments that have moved the humanities platial sandbox beyond a two-dimensional framework. Collectively, the volumes demonstrate how deep maps that offer a rich and deep experience of place can be a support for the scholarly imagination, both for the construction of an argument and for the presentation of a narrative. Deep maps are now a valid and valued method of scholarship and can be a basis for understanding both pragmatic and scholarly needs.

Finally, the series suggests there is, indeed, a basis for collaboration between GIScientists and humanities scholars and that this basis can be found in the progress the GIScience community has made since the GIS in Society critiques of the mid-1990s. These efforts include participatory GIS, qualitative GIS, neogeography and volunteered geographic information, and critical GIS, all approaches that are mentioned within these volumes and that continue to be active areas of research and discussion within the GIScience community.[13] There is hope that the spatial humanities and this series bring even more imperative to these issues.

What is now possible?

What is next for deep maps and the spatial humanities? These three volumes provide a fertile field for anyone wishing to commit to a life's work on a deep map. How this might be done is hinted at in some examples in the series. How all of these insights might be engaged to produce a deep map are yet to be discovered but are limited only by the imagination of the scholars who want to move to the next level of deep mapping.

There are many new technologies ripe for exploitation by those who are interested in deep mapping. Trevor Harris' examples of how place can be experienced using immersive technologies point at the possibilities. Immersive technology is an entirely new medium, a light year away from a flat two-dimensional map, and it carries with it the opportunity to take advantage of perceptual affordances not previously generally available to a scholar or the public. If a deep map is intended to provide opportunities to experience places in all their varied details and complexities, the immersive environment allows that kind of tool, merging GIS-mediated space with place. Within the immersive environment, all of the place-based experiences possible do not need to be precast or selected. Just putting the observer into the place allows all these new senses and perceptions to be experienced.

The GIScience community continues to engage actively with many of the themes raised in these volumes. The focus of the 2019 annual symposium of the University Consortium for Geographic Information Science (UCGIS) was "The Geospatial Humanities," under the co-chairmanship of the then UCGIS president, Alberto Giordano, the editor of a related publication in the Series of Spatial Humanities (Indiana University Press) in which the three volumes under discussion here appear. Several papers from the symposium were collected into a special double issue of the *International Journal of Humanities & Arts Computing* (Vol. 14, Issue 1/2, March 2020). These highlight ways in which the GIScience community is working to make some of the necessary technological advancements called for in this series to enable the envisioned deep maps.

Big data (both spatial data and digital archives) accessible in a semantic web paired with machine learning provide additional opportunities for new deep map environments and scholarly discoveries. Using machine learning techniques to uncover spatial narratives is already being done, though mostly in pragmatic fields like geo-business and geospatial intelligence. Coupling that technology with the exploratory, multifaceted, deep contingency context of the humanities suggests the opportunity to uncover unlimited spatial narratives. Machine learning needs seeds of some sort, such as ontologies, key words, targeted place/time spaces. These seeds set the context of a narrative and different narratives will arise from different seeds.

There is a bold, bright future for deep maps in the spatial humanities and exciting opportunities in GIScience with the emergence of new digital environments and methods that allow us to escape the confines of Cartesian space and structured data layers.

Notes

1 Introduction, 5, in David J. Bodenhamer, John Corrigan, and Trevor M. Harris, eds., *Deep Maps and Spatial Narratives*. Series on Spatial Humanities (Bloomington, IN: Indiana University Press, 2015).
2 David J. Bodenhamer, "Narrating Space and Place," in Bodenhamer, et al., *Deep Maps and Spatial Narratives*, 10.
3 May Yuan, "Mapping Text," in David J. Bodenhamer, John Corrigan, and Trevor M. Harris, eds., *The Spatial Humanities: GIS and the Future of Humanities Scholarship*. Series on Spatial Humanities (Bloomington, IN: Indiana University Press, 2010).
4 Bodenhamer, "Narrating Space and Place," in Bodenhamer, et al., *Deep Maps and Spatial Narratives*, 14.
5 Trevor M. Harris, John Corrigan, and David J. Bodenhamer, "Challenges for the Spatial Humanities: Toward a Research Agenda," in Bodenhamer, et al., *The Spatial Humanities*, 173–4.
6 May Yuan, John McIntosh, and Grant DeLozier, "GIS as a Narrative Generation Platform," in Bodenhamer, et al., *Deep Maps and Spatial Narratives*, 179.
7 John Corrigan, "Qualitative GIS and Emergent Semantics," in Bodenhamer, et al., *The Spatial Humanities*, 81.
8 David J. Bodenhamer, "The Potential of Spatial Humanities," in Bodenhamer, et al., *The Spatial Humanities*, 29.
9 Trevor M. Harris, John Corrigan, and David J. Bodenhamer, "Conclusion: Engaging Deep Maps," in Bodenhamer, et al., *Deep Maps and Spatial Narratives*, 224.
10 John Corrigan, "The Inexactitude of Science: Deep Mapping and Scholarship," in David J. Bodenhamer, John Corrigan, and Trevor M. Harris, eds., *Making Deep Maps: Foundations, Methods, and Approaches*.
11 Trevor M. Harris, "Deep Mapping the Lived World: Immersive Geographies, Agency, and the Virtual Umwelt," in Bodenhamer, et al., *Making Deep Maps*.
12 Stephen Robertson and Lincoln A. Mullen, "Navigating through Narrative," in Bodenhamer, et al, *Making Deep Maps*.
13 S. Warren, T. Harris, M. Goodchild, and P. Solis, "Commentaries on Evaluating the Geographic in GIS," *Geographical Review* 109, no. 3 (2019): 308–20.

Index

Note: Page numbers in *italics* indicate a figure.

Adkins, Brent 162
Anawalt, Patricia Rieff 93
Annales 2
Anthropocene epoch 38, 40, 42, 45
Arcades Project, The (Benjamin) 67
argument 45–6
assemblages 40–1, 43, 162–6, 172
Ayers, Edward 133, 178

Bakhtin, Mikhail 6, 10, 72
Bauch, Nicholas 9, 125
Beck, Robert J. 28
Beckett, Samuel 65
Bellamy, John 21
Benjamin, Walter 67
Bennett, Jane 38, 42, 47
Bertron, Cara 20
Blair Mountain 122–5
Bloomsday 71, 74
Bodenhamer, David J. 50, 54, 149
Borchardt, Rudolf 67
Borges, Jorge Luis 162
Borgia 84, *85*
Bowden, Martyn 19
Brattleboro Museum and Art Center 17–19
British Library 149–51, 153, 155, 158–9
Buchin, Kevin 56
Buttimer, Anne 65, 66

Campos, Ricardo 52
cartography *see* maps and mapping
Casey, Edward 2
Caso, Alfonso *95*
Casti, E. 114
Çatalhöyük Map 1
cataloguing 152–4, 156
Ce Acatl Quetzalcoatl 85

Certeau, Michel de 2
Chaplin, Ralph 123–4
chronotope 6–7
*Circulation of Knowledge and Learned
  Practices in the 17th Century Dutch
  Republic* 10
Clement, Tanya 52
Cliodynamics 170
*Códice de Santa María Nativitas de
  Atengo* 92, 93
collages 41–3
copyright 154–5
Corrigan, John 12, 149
Cortés, Hernán 79
Crace Collection of Maps of London
  151–2
Crace, Frederick 152
Crampton, J. W. 115
Crawford, Carter 27
Cronin, Anthony 71
crowdsourcing 169
Cultural Atlas of Australia 8

Dante Alighieri 71–2
Danteworlds 170–1
Darnton, Robert 133
Davis, Rebecca Harding 11, 120–2, 125
de Botton, Alain 21
De Certeau, M. 112, 117
Debord, Guy 44, 46, 112
Deconstructing the Map (Harley) 115–16
Deconstructionism 2
Deely, J. 118
deep mapping: assessing 171–2; and
  cartography 75, 114–16, 119, 127–8;
  categories of 7–10; conceptualizing
  162–3; definition 6–7, 50, 148–9;

experiential science 32–4; and fieldwork 28–31; and historic collections 149; macro scale 38–40; and memories 21; meso scale 40–3; methodologies 112–13; micro scale 43–5; origins 1–2, 19; principles of 7; purpose of 50, 78, 113–14, 132, 134–5, 177–8; and stories 17–19; structure of deep maps 51; vs. "thin" maps 166; value of 12–13, 106; *see also* maps and mapping
Deep Mapping and the Spatial Humanities (Bodenhamer et al.) 53, 54
Deep Maps and Spatial Narratives 175
Deleuze, Giles 40, 43, 162–3
Dewsbury, J. D. 41, 43, 116–17
digital data *see* technology
Digital Public Library of America (DPLA) 8
Dodge, Martin 75
Dourish, P. 116
Downs, Gregory 136–7
Dyer, Geoff 21

Ebstorf Map 1–2
Eco, Umberto 39
Eliasson, Olafur 21
emotions 163–4
Enchanting the Desert (Bauch) 9
Enchantment of Modern Life, The (Bennett) 47
encounter 40, 42–6
Enlightenment 2
environmental conditions 125–7
environmentalism 40–1, 125–7
Epilegomena zu Dante (Borchardt) 67
Escalante Gonzalbo, Pablo 102, 107n6
EVE Online 11–12
Everything Sings (Wood) 28, 29, *32*, *33*, *35*, *36*

Fernández-Christlieb, Federico 85
Finlay, Ian Hamilton 32
Fisher, Shelly Fishkin 168
Fleeting Glimpses (Wood) 22, *24*
Flemming, William 18
Fogli, D. 117
Foucault, Michel 3

Gaddis, John Lewis 74
gazetteers 2
Geographies of the Holocaust 164
geography 3, 40, 66
geoparsing 156–7
Georeferencer 158

George, Bob 18
geosophy 19–20, 65–76
geotagging 55–60
Giaccardi, E. 117
Giordano, Alberto 179
GIS (Geographic Information Systems): extracting information 51; German Historical GIS 8; Great Britain Historical GIS 9; and humanities 133–4, 175–80; HumGIS 66–7, *70*, *71*, 75; impact of 113; in Latin America 105; purpose of 3–4, 78, 127–8; use of 5
Glass, Ira 28
Golding, William 75
Goodchild, Michael 75
Google Arts and Culture 8
Google Books 50–1
Gray, Andrew 159
Great Britain Historical GIS 9
Gregory, Ian 157, 163
Gruzinski, Serge 97
Guattari, Felix 40, 162–3

Hägerstrand, Torsten 11
Hall, Becca 20
Harlem 138–41
Harley, J. B. 115–16
Harman, Graham 40
Harris, Trevor M. 11, 65, 133, 149, 178, 179
Harvey, David 3
Hawkins, Harriet 40
Heat-Moon, William Least 1, 19, 112
Heeney, Robert Patrick 69
Heidegger, Martin 2, 6
Hey, T. 165
Hill Boone, Elizabeth 86
Home Rules (Wood and Beck) 28, *29*, *30*, *31*
humanism 132
humanities: digital 44–5, 132–3, 170; spatial 78, 105, 133, 163, 166, 170–1, 175–80
Husserl, Edmund 6, 112
HyperCities 5

I Don't Want To But I Will (Wood) 25
Indigenous traditions: conception of cosmos 81–2; decline of 102–3; mapping 80–1, 99–100; mnemonic technologies 87–90, *89*; spatial conceptions 82–6, *83*, *84*, *85*, *86*; and value of deep mapping 106; writing 86–7, 92, 94, *99*, *101*; *see also* New Spain and Mesoamerica
Inferno (Dante) 71–2

IN-SPIRE *56*
International Journal of Humanities &
Arts Computing 179
Ireland, Republic of 67–74

James, William 12, 169
Joyce, James 67, 71–2

Kaski, Samuel 54
Kavanagh, Patrick 71
Kemp, Karen 175
Kimble, George H. T. 46
Kings Topographical Collection 154,
158–9
Kitchin, Rob 167
Kohonen, Teuvo 55
König, Viola 99
Korzybski, Alfred 172n2
Kuhn, Thomas 165
Kundera, Milan 66

Latour, Bruno 167, 169
Lefebvre, Henri 4
Leonardo da Vinci 75
León-Portilla, Miguel 88
Li, Dongying 56
libraries 149–54
Lienzo de Petlacala 103–4, *104*, *105*
Life in the Iron Mills (Davis) 11
lifepaths 69
lighting 125–7
López de Velasco, Juan 79–80, 102, 107n1
Lopez, B. H. 112
Lorimer, H. 117
Lovelock, James 75

Map Communication Model (MCM) 115–16
Mapping Emotions in Victorian London
163–4, 168
Mapping Medieval Chester 9
Mapping Occupation 136–7, *137*
maps and mapping: affective 117–18; as
art 25–7, *27*; as assemblages 162–3;
cognitive maps 20, 22; collections
151–2; commodification of 4; and
deep mapping 75, 114–16, 119, 127–8;
definition 50; digital 167–8; Indigenous
vs. European styles 80–1; modalities
25; purpose of 62; techniques 28; thick
mapping 5; uses of 20; *see also* deep
mapping
Massey, Doreen 2, 9
mathematics 173n5
Mercator, Gerardus 151

Mesoamerica *see* New Spain and
Mesoamerica
Mignolo, Walter 86
Misrach, Richard 40
Monmonier, M. 115
Moore, Robin 25
Moretti, Franco 6
Moulahi, Bilel 52
Mullen, Lincoln A. 177–8
Mundy, Barbara 99

Named Entity Recognition (NER) 58
Natural History and Antiquities of
Selbourne (White) 19
Neatline 136–44, 147
Nesbit, Scott 136–7
New Spain and Mesoamerica: formation
of 79–80; mapping of 80, 88–102, *89*,
91, *95*, *96*, *97*, *98*, *100*; written tradition
86–7; *see also* Indigenous traditions
Nietzsche, Friedrich 3
nonrepresentational theory 116–17
November, Valérie 167, 169
Nowotny, Karl 87

O'Brien, Flann 71
Oculus Rift 5
Olko, Justyna 92
Olsson, Gunnar 74
Olvera, Manuel de 92, 93
Overlook 20–2, *21*, *22*

Padilla, Juan de 90, 93
Pearson, M. 112–13
Pearson, Michael 2, 112–13
Pelagios Commons 10
Picasso, Pablo 41–2
place 116
Places Journal 29
Pocket Guide: to the known world 20
Poole, Nick 150
Popper, Karl 165
Porras, Don Martin de 92
Portrait, The 17, *18*, 19
Postmodernism 2–4
PrairyErth (Heat-Moon) 1, 19
Presner, Todd 9
Ptolemy 1

Quercia, Daniele 163

Rashomon 68
Relaciones Geográficas del siglo XVI
79–80, 97, 102–3

RICHES 8
Robertson, Stephen 136, 138, 177–8
Rocky Mountain Arsenal National Wildlife
    Refuge 40–1
Rubin, Anne Sarah 133
Russo, Alessandra 94

Salem Witch Trials 9
Salling, Mark 21, *23*
Santa Cruz, Alonso de 79
scholarship 166–72, 176–8
science 164–7
Self-Organizing Maps (SOMs) 54–5, 61
Sewell, William 12
Shanks, Michael 2, 112–13
Simpson, P. 117
Situationists 1
Sloane, Hans 149
Smith, Mary Elizabeth 99
Smith, Rachel 163
Society of the Spectacle, The (Debord) 46
Soja, Edward 3
space 116
Spatial Humanities, The 175
spatializing text: in geographic spaces
    55–60, *56*, *59*; purpose of 50, 60; in
    semantic spaces 52–5, *53*, *54*
spatial narrative 134–44
spatial turn 2–4, 113
Stegner, Wallace 19
stereoscopes 67, 74
Strabo 66
Straughan, Elizabeth 40

technology: application of 170; availability
    of texts 50–1; convergence of
    4–5; digital data 148–52, 154–60; and
    narrative 136–44; and word clouds 52–5
"*Terrae Incognitae*: The Place of
    Imagination in Geography" (Wright)
    19–20
Thomas III, William 133

Thoreau, Henry David 19
TimeMap 178
Tlacuilolli (Nowotny) 87
Tolle, K. 165
toponyms 57–9, 61, 82, *101*
topophilia 2
Torquemada, Juan de 87
Troubled Geographies 164
Tuan, Yi Fu 116
Turchin, Peter 170
Twitter 57

Uexküll, Jakob von 118
Ulysses (Joyce) 67–8, 71–4, *72*, *73*
umwelt 116–20

Valley of the Shadow 8–9, 133, 178
Vico, Giambattisto 65
virtual reality 11, 119–27, *121*, *122*, *124*,
    125, *126*, *127*
Vision of Britain 9
visual representation 46–7

Walden (Thoreau) 19
Warf, Barney 3
Weber, Max 38, 42–3
Wheeling, West Virginia 120–2, *121*, *122*
Where You Are 21, *23*
White, Gilbert 19
Wikipedia 151
Wolf Willow (Wallace) 19
Wood, Chris 21
Wood, Denis 17, 22, *24*, 25, 163, 178
word clouds 52–60, *53*, *54*
Wright Carr, David Charles 90
Wright, J. K. 19–20, 65

Xiaolu Zhou 56

Year of the Riot 138–44, *139*, *141*, *142*
Yeats, William Butler 68
Yiu Fu Tuan 2

Printed in the United States
by Baker & Taylor Publisher Services

Printed in the United States
by Baker & Taylor Publisher Services